THIRD EDITION

Anatomy & Dissection of the Rat

Warren F. Walker, Jr.
PROFESSOR EMERITUS, OBERLIN COLLEGE

Dominique G. Homberger
PROFESSOR, LOUISIANA STATE UNIVERSITY, BATON ROUGE

ILLUSTRATED BY

Fine Line Illustrations

W.H. FREEMAN AND COMPANY
NEW YORK

First printing 1997, QP

TABLE OF CONTENTS

v

TEACHER'S

GUIDE

ONLY A FEW REMARKS are necessary for a set of anatomical exercises of this type. They relate to the specimens and their care, to useful supplemental materials, and to particular parts of the exercises in which students might need special help.

The rat is a relatively small animal with accordingly small organs and structures. Some structures, such as certain salivary glands, some muscle, nerves of the autonomic nervous system, and smaller blood vessels are especially difficult to observe with the naked eye of even young students. A low-cost alternative to making dissecting microscopes available to students is magnifying lamps (3×). Various models are available at office supply and other stores for a relatively modest cost. Several of these lamps for demonstration specimens and for students to use for the smallest structures can make a big difference in the student's comprehension of the described materials.

Exercise 1
External Anatomy, Skin, and Skeleton

Preserved specimens of rats can be obtained from any biological supply company. Large specimens should be ordered. If the students are to dissect veins as well as arteries, they should work on doubly injected specimens. You can save some time and money by giving the students singly injected specimens, in which the arteries are injected, and provide demonstration dissections of the veins.

Specimens usually are fixed in formalin (an aqueous solution of formaldehyde gas), because it penetrates tissues rapidly and chemically links proteins, thereby protecting them from enzymatic and bacterial decay. After fixation, specimens used to be stored in formalin, but now they usually are transferred to a holding solution based on isopropanol, ethylene glycol, or a 1–2% phenoxyethanol solution. These solutions prevent mold and tissue deterioration. Isopropanol, however, is very flammable, so sparks and open flames should be avoided in its presence. As specimens are dissected, they should be sprayed or doused with a wetting solution, such as a 1% phe-

noxyethanol solution. Phenoxyethanol is a nontoxic preserving agent with a rather pleasant scent, and it softens and rehydrates the tissues of the specimen. To prepare such a solution, mix 1 part of phenoxyethanol (2-phenoxyethanol practical grade, from Eastman Chemical Co.) with 99 parts of very hot tap water. Because phenoxyethanol is slightly lipophilic, you may wish to stir the solution on a magnetic stirrer for about 30 minutes to ensure a homogeneous solution.

Between laboratory sessions, wrap the specimens in cloths soaked with the phenoxyethanol solution and store them in tightly closed plastic bags or boxes. If the specimens become hard and dry, add 1 part of any laundry fabric softener to 99 parts of the 1% phenoxyethanol solution and immerse the specimens in this. Additional information on this and other techniques can be found in the following references, which are cited in full in the References: Blaney and Johnson (1989), Frolich et al. (1984), Rosenberg (1992 a and b), and Wineski and English (1989). If specimens are held in solutions provided by the supplier, follow the supplier's instructions with respect to their care.

B. Skin

Descriptions of microscopic anatomy throughout these exercises are based on histological slides of mammalian tissue, usually from a cat or a rat. The differences between tissues of these animals are negligible at the level at which the organs are being studied. Individual histological slides may be provided to the students, or a series of demonstrations may be set up. If students are given individual slides, some structures may not be present in all of them. Therefore, certain structures should be shown in demonstration preparations: sweat gland, capillary, and arrector pili muscles.

C. Skeleton

A mounted rat skeleton and human skeleton should be available as well as separate human skulls, and individual human thoracic or lumbar vertebrae. It is easier to study these structures on human specimens than in the rather small rat material.

D. Skeletal Materials

Time can be saved if students study cartilage on demonstration histological slides and concentrate on bone. Demonstrate a bone cut longitudinally so that students can see the difference between compact and spongy bone. Slides of dried compact bone, thinly ground, show bone structure more clearly than slides of fixed bone.

Exercise 2 Muscles

D. – J. Muscle Groups

Muscles are described by regions of the body so that instructors can select certain groups if time does not permit a complete dissection. It is necessary to dissect the shoulder muscles before those on the cranial part of the trunk can be studied, but other groups can be studied independently.

K. Muscle Tissue

If histological slides of muscle tissue are not available, the tissues can be seen in slides of certain organs: tongue for skeletal muscle, heart for cardiac muscle, and intestine for smooth muscle. Demonstrate skeletal muscle on a slide in which some of the individual muscle fibers are separated. A high-power, oil-emersion demonstration is needed to show A and I bands clearly.

Exercise 3 Digestive and Respiratory Systems

A.1. Salivary Glands

Because salivary glands are very difficult to dissect, an instructor may wish to have students study them on a demonstration dissection. Even if students do their own dissections, a demonstration will help them find the glands.

A.2. Mouth, Pharynx, Larynx, and Neck

Demonstrating to groups of students the procedure for opening the mouth and pharynx will save a lot of time and prevent destruction of some of the structures in this region.

Certain features of the head and neck can be seen particularly well in sagittal sections, and several such sections should be on demonstration.

Vocal cords are very small in a rat, but can be seen on a demonstration of the cow's larynx. If a fresh larynx and trachea is obtained from a local slaughter house, a "moo" can be produced by blowing through the trachea and manipulating the larynx.

B. Opening the Body Cavity

Time can be saved by demonstrating to groups of students the procedure for opening the body cavity.

Exercise 4 Circulatory System

E. Root of the Lung and Internal Structure of the Heart

The heart of the rat is large enough to show most parts of the mammalian heart clearly, but it is helpful to have a demonstration dissection of a larger sheep heart.

If you can obtain a fresh sheep or beef pluck (heart and lungs removed as a unit) from a local slaughter house, you can simulate with water the flow of blood through the heart, and the action of the heart valves in preventing backflow. An educational institution is usually given permission to obtain a fresh pluck in which the heart has not been cut open. Slaughter houses are normally required to slice open the heart to drain blood from it.

Dissect away the pericardial wall and other tissues to expose the major blood vessels entering and leaving the heart. Look for and save the ligamentum arteriosum connecting the pulmonary trunk and aortic arch. Cut open the right atrium and left ventricle, and remove any clotted blood. Also nick one pulmonary artery and one pulmonary vein.

Dissect very carefully on the left and right side of the base of the aorta to find the origin of the coronary arteries. Because the left ventricle has been cut open, the coronary arteries will leak when water is inserted into the aorta. Tie or clamp off the right coronary artery, and slip a piece of cloth beneath the left coronary artery so it can be tied off when necessary. Insert a hose line from a water tap into the brachiocephalic artery, tie it securely in place, and tie off the distal end of the aorta. A second hose line, which can be taken off the first hose line by a Y-tube, should be available for inserting into various structures. A clamp will be needed to direct water into one or the other hoses.

To simulate the flow of oxygen-depleted blood into the lungs, insert the free hose through the right atrium into the right ventricle. Water (oxygen-depleted blood) will emerge through the cut pulmonary artery. Clamp off the pulmonary trunk with your hand to increase the water pressure in the right ventricle; the closed right atrioventricular valve can be seen by looking into the right atrium.

To simulate the return of oxygen-rich blood from the lungs, insert the free hose into the cut pulmonary vein and observe water entering the cut left ventricle through the left atrioventricular valve.

The action of the aortic valve can be demonstrated by running water into the hose that has been tied into the brachiocephalic artery. Clamp off the left coronary artery when you do this. The aorta will swell as pressure builds up, and the closed aortic valve can be seen by looking into the left ventricle at the point of origin of the aorta. While there is still water pressure in the aorta, release the left coronary artery and water will spurt through its branches, which were cut in opening the left ventricle. This demonstrates the separate coronary circulation.

F. Blood

If students are given slides or smears of mammalian blood to study, demonstrations, preferably under an oil-emersion objective, of some of the leukocytes should be available. Eosinophils and basophils are particularly hard to find.

Exercise 5 Urogenital System
A. Excretory System

Sheep or pig kidneys in which the arteries, veins, and ureter have been injected can be obtained from biological supply houses. They make excellent demonstrations. Even the location of glomeruli in the cortex can be seen with low magnification.

B.2. Female Reproductive Organs

Pregnant rats, or the uterus of a pregnant sow, can be obtained from biological supply houses. Corpora lutea in the ovary, placental structure, fetal membranes, and embryos can be demonstrated on them.

C. Gonads

Give each student histological slides of the mammalian testis or ovary, or show these structures on demonstration slides.

Exercise 6 Nervous Coordination: Nervous System
B. – F. Meninges to Brain Structure and Functions

The meninges and major regions and structures of the mammalian brain can be seen quite well on a rat brain, but it is rather small, so demonstrations of a larger sheep brain should be available. Descriptions of a sheep brain are available as a Separate (889) from W. H. Freeman and Company. The following have proved to be helpful demonstrations: the brain with the dura mater intact, a dissection in which the cranial nerves have been carefully exposed, and sagittal sections of the brain.

G. Spinal Cord, Spinal Nerves, and Autonomic Nervous System

It is difficult to dissect the spinal cord in an animal as small as the rat, but their basic structure shows very well in histological slides through these structures in a frog or a mammal. Frogs are small enough that slides made from them usually show the spinal cord and a spinal nerve in the same preparation.

Exercise 7 Nervous Coordination: Sense Organs
A. Eye

The eyeball of a rat is small and difficult to dissect, so a demonstration dissection of the larger eye of a sheep or cow should be available. The directions for dissection apply to most mammal eyes.

Teacher's Guide

D R. DOMINIQUE G. HOMBERGER of Louisiana State University, Baton Rouge, has joined the senior author in preparing the third edition of this laboratory manual. The manual is enriched by the perspectives of two of us, and the participation of an active teacher helps meet the needs of current students. We have prepared this edition, like its predecessors, primarily to meet the need for a short set of exercises appropriate for college students, or advanced high school students, studying mammalian anatomy with reasonable thoroughness in introductory zoology or biology courses. Although the animal studied is a rat, the exercises also apply to the mouse. Because rats and mice are mammals, their anatomy reflects that of other mammals, including human beings, and we have indicated major points where their anatomy differs from ours. Rats are often used to exemplify mammals because rat specimens are abundant, inexpensive, and easy to store. Although not as large as a fetal pig or a cat, they have surprisingly large internal organs that are easy to study. Rats and mice are also the most widely used mammals for experimental purposes in colleges and universities. Because a large number of students, including some of those engaged in experimental work, are ignorant of the anatomy of these important animals, some of these exercises may be of use to students in physiology and other advanced courses.

We have also included descriptions of a few sheep organs, such as the brain and eye, because they are larger and much easier to study than the comparable organ of the rat—although the rat brain is considered. Each exercise deals with a natural unit of material, usually one or two organ systems. Gross anatomy is emphasized, but the functions of the various parts and their functional interrelationships with other parts are also stressed. Gross structure cannot be understood without these perspectives. Directions for dissection are thorough enough to enable students to proceed with a minimal amount of help from the instructor and to encourage them to take more than a superficial look at the material. Careful observations are essential for verification of descriptions: they should enhance the student's understanding of both the potentialities and the limitations of the descriptive

aspects of science. In addition, they are necessary for evaluating the generalizations that are derived from an anatomical study.

Although the emphasis is on gross anatomy, directions for the study of the microscopic anatomy of selected tissues and organs are included. A knowledge of the microscopic structures of organs adds greatly to an understanding of their gross structures and functions, and such material is presented frequently in introductory courses.

Enough material is included so that these exercises also meet the needs of students in some comparative and mammalian anatomy courses. We have aimed for a flexible presentation, because these exercises are used in many different types of courses. If an instructor plans to omit exercises, he or she may wish to order the individual exercises desired rather than work from this entire set. Furthermore, parts of individual exercises can be replaced by demonstrations or omitted entirely, which gives the instructor the flexibility to adjust the material to meet his or her specific needs.

The drawings are intended to aid students in finding the essential parts and to afford a permanent reference to the structures observed in dissection, after the specimens have been discarded. Many of them show structures in relation to surrounding parts; they not only make it easier to identify the structures, but also emphasize the unity of the organism. If time permits, students should be encouraged to make additional sketches, because very careful dissection and close observation are required before an organ can be drawn accurately.

Favored technical terms are printed in boldface type in the exercises themselves. For the most part, they are anglicized versions of those recommended by the *Nomina Anatomica Veterinaria*. This code, which applies, insofar as possible, the human terms of the *Nomina Anatomica* to quadrupeds, is becoming the standard for mammals other than human beings. In both codes, the terms describe some aspect of the organ. Accordingly, the use of eponyms (terms named after a person) is avoided. We follow this convention, but give the eponym as a synonym if it is familiar to many people, such as uterine tube (Fallopian tube) and **osteon** (Haversian system). The two codes differ primarily with respect to modifying adjectives for direction: the human "superior vena cava," for example, is referred to as the "cranial vena cava." In a few cases we use a familiar English word instead of the less familiar *Nomina* term (e.g., "liver" rather than "hepar"); however, in such cases, the *Nomina Anatomica* term is given in *italics* because it often forms the basis for constructing familiar adjectives (e.g., "hepatic"). A few other common synonyms are given in roman type.

We have been guided in preparing the third edition by comments sent to the publisher by users of the manual, and by reviews received from the following teachers: Brenda J. Claiborne, University of Texas at San Antonio; Prentiss G. Cox, Mississippi College; James G. Fox, Massachusetts Institute of Technology; Gwilym S. Jones, Northeastern University; R. David Jones, Adelphi University; and Barry Thomas, California State University at Fullerton. The suggestions of these instructors have been helpful, and we are grateful to them. For various reasons, however, we could not incorporate all of the suggestions made, and the reviewers are not responsible for any errors or shortcomings that remain. We hope users of this edition will continue to call our attention to any errors or parts in need of revision.

We have reworded and modified sections in this edition that appeared to be unclear and difficult for students. We have expanded or added sections that describe the functions of the organs being studied. This new material both reduces the laboratory time needed for instructors to explain the significance of the material and reinforces lecture material. Many figures have been modified and relabeled to make them clearer, others have been added, and many photomicrographs of microscopic anatomy are included. The **Glossary of Vertebrate Anatomical Terms** has been completely redone. We now include a phonetic pronunciation for each term as well as its classic derivation and meaning. If students study the derivations of terms, they will soon recognize that many roots are repeated in different combinations and the meaning of unfamiliar terms will become easier to learn. We have included a table of contents when these exercises are bound as a complete set.

Acknowledgments

Fine Line Illustrations has redrawn electronically all of the art for this edition. This has allowed us to update the drawings and add new figures, but we continue to be indebted to the original artists for the great care and artistry with which they prepared the original drawings for previous editions: Judith L. Dohm, Edna Indritz Steadman, Pat Densmore, and Tom Moore. Editorial Services of New England has contributed a new and rejuvenated cover and interior design by John Svatek, and art and editorial management by Sarah Kimnach. We are also much indebted to Jodi Simpson, whose editing of these exercises has eliminated many ambiguities and errors, and to the outstanding staff at W. H. Freeman and Company who guides us through the many stages of writing and production: Erica Seifert, Project Editor; Patrick Shriner, Sponsoring Editor; Melina LaMorticella, Assistant Editor; and Maura Studley, Production Supervisor.

Warren F. Walker, Jr.
Ossipee, New Hampshire, 1997

Dominique G. Homberger
Baton Rouge, Louisiana, 1997

STUDENT'S

GUIDE

F THIS IS THE FIRST ANIMAL that you intend to dissect, you should read these remarks concerning procedures and terms of direction carefully before you begin.

A dissection is to expose you to important organs, to display their spacial relationships to surrounding ones without destroying them, and to find where the organs begin and end. If you perform a dissection well, another person should be able to examine the specimen and see clearly what the organs are, where they lie, and how far they extend.

A few cuts, or **incisions,** must be made at the outset with a scalpel or a pair of scissors to open the specimen, but thereafter most of the dissections should be done with two pairs of fine forceps, one to hold up an organ and the other to pick tissue away from it. Dissecting is done by carefully separating organs and picking away surrounding extraneous connective tissues, such as loose connective tissue, superficial fascia beneath the skin, and fat, or adipose, tissue. As you continue to dissect, you sometimes will be asked to **bisect** a structure; that is, to cut across its center and to **reflect,** or turn back, its two ends to expose an underlying organ. Do not remove any organs unless specifically directed to do so, but always pick away enough of the surrounding tissue to see all of an organ clearly. Do not be content with just a glimpse of an organ.

During the dissection, you may want to spread the legs of the specimen and tie them to the dissecting pan, but it is not necessary to do this.

The terms **right** and **left** always refer to the *specimen's* right and left sides. Depending on how the specimen is oriented, this may or may not correspond to your own right and left. **Lateral** refers to a direction toward the side of the body or organ in question; **medial,** to a direction toward the center of the body. When a structure is described as being **superficial** to some part, it means that it lies over the part referred to and nearer to the body surface. Conversely, a structure that is **deep** to some other structure, lies beneath it and farther away from the body surface. Other terms for direction differ somewhat for a quadruped and for human beings, who stand erect (see the accompanying figure). For human beings, **superior**

refers to the upper, or head, end of the body; **inferior,** to the lower parts of the body. The belly surface is **anterior** and the back is **posterior.** For quadrupeds, the underside of the body is **ventral** and the back is **dorsal.** Although the terms *anterior* and *posterior* are sometimes used to refer to directions toward the head and tail, respectively, in quadrupeds, many anatomists avoid using them in a quadruped because of their different use in a biped. **Cranial** describes a direction toward the head; **caudal** describes a direction toward the tail. **Rostral** describes the direction within the head toward the snout. **Anterior** and **posterior** are used in describing some individual organs, such as an eyeball, where the orientation is essentially the same in human beings and quadrupeds.

The **distal** end of a structure is the end farthest away from some point of reference, usually the origin of the structure or the midventral line of the body; the **proximal** end is the end nearest the point of reference.

A **sagittal section** is a section in the longitudinal plane of the body passing from the middle of the back to the middle of the belly. A **frontal section** is also a longitudinal section, but it crosses the sagittal plane at right angles. A **transverse section** crosses the longitudinal axis of the body at right angles.

The organs and structures you expose have names. Learn them so that you and your teacher can communicate effectively, but try not to be overwhelmed by them. Most terms are anglicized versions of Greek and Latin words that describe some aspect of an organ's appearance, location, structure, or function. For example, **acetabulum,** the socket in the hip bone for the femur, is a Latin word meaning "vinegar cup" (acetic acid has the same root); **carpals,** the small bones in the wrist, comes from the Greek *karpos* meaning "wrist"; the muscle action **abduction** comes from two Latin words, *ab-* and *ductus* meaning "leading away." If you occasionally look up the derivations and meanings of terms in the Glossary, you will soon become familiar with the classical roots, many of which are used repetitively, and you will learn additional anatomical terms more easily.

At the end of a laboratory period, always clean up your work area and put your specimen away in the container provided.

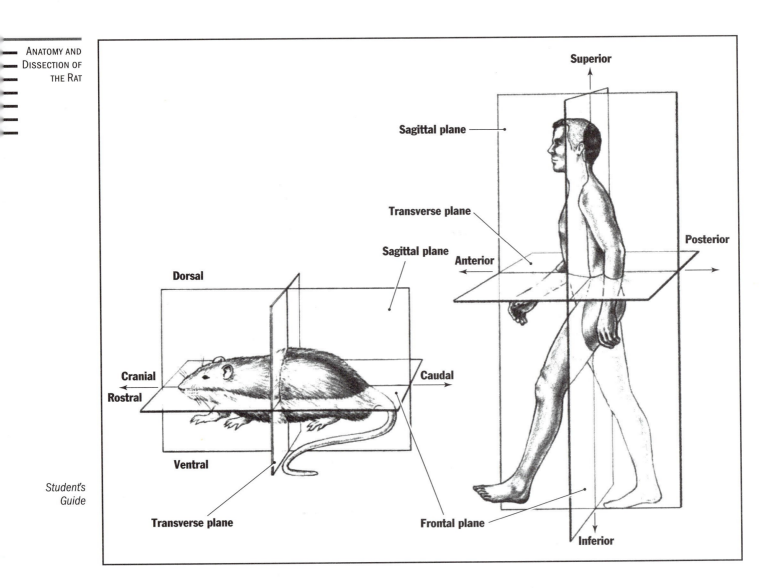

Superior

Sagittal plane

Transverse plane

Sagittal plane

Anterior

Posterior

Dorsal

Cranial

Rostral

Caudal

Ventral

Inferior

Transverse plane

Frontal plane

References

The following list of references will be of use to those who wish to pursue various aspects of the biology of rats and mice, or of mammals in general. The list is not intended to be exhaustive; it is an introduction to the extensive literature that is available and includes only references that are particularly useful and a few special ones that have been cited by the authors.

Blaney, S. P. A., and Johnson, B. *Technique for reconstituting fixed cadaveric tissues.* Anatomical Record 224: 550–551, 1989.

Chiasson, R. B. *Laboratory Anatomy of the White Rat,* 5th ed. Dubuque, IA: Wm. C. Brown, 1988.
  A standard laboratory manual.

Cook, M. J. *The Anatomy of the Laboratory Mouse.* New York: Academic Press, 1965.
  A set of line drawings of the anatomy of the mouse. It is very useful for those who wish to compare rat and mouse structures.

Dorit, R. L., Walker, W. F., Jr., Barnes, R. D. *Zoology.* Philadelphia: Saunders College Publishing, 1991.

Fawcett, D. W. *Bloom and Fawcett a Textbook of Histology,* 12th ed. New York: Chapman & Hall, 1994.
  Although based primarily on human histology, this book is a valuable reference for all aspects of mammalian histology and cytology.

Frewein J., Habel, R. E., and Sack, W. O. *Nomina Anatomica Veterinaria,* 4th ed. Cornell University, Ithaca, NY: World Association of Veterinary Anatomists, 1994.
  A list of standardized Latin names for organs and anatomical structures for mammals.

Frolich, K. D., Anderson, L. M., Knutsen, A., and Flood, P. R. *Phenoxyethanol as a nontoxic substitute for formaldehyde in long-term preservation of human anatomical specimens for dissection and demonstration purposes.* Anatomical Record 208: 271–278, 1984.

Green, E. L., Ed. *Biology of the Laboratory Mouse,* 2nd ed. New York: McGraw-Hill, 1966.

This important book, written by the staff of the Jackson Laboratory, covers nearly all aspects of the biology of the mouse: husbandry, genetics, reproduction, embryology, anatomy.

Greene, E. G. *Anatomy of the Rat.* New York: Hafner, 1963.

This book was originally published as a Transaction of the American Philosophical Society, New Series, Vol. 27, 1935. It remains the most comprehensive study of the anatomy of the albino rat.

Hebel, R., and Stromberg, M. W. *Anatomy and Embryology of the Laboratory Rat.* Wörthsee, Germany: Biomed Verlag, 1986.

Thorough descriptions of the gross, microscopic, and embryonic anatomy of the rat are presented primarily for the research scientist, but the book will be of value to all students of rat anatomy. The illustrations are the best available. Hundreds of references are given to the primary literature.

Junqueira, L. C., Carneiro, J., and Kelley, R. O. *Basic Histology,* 7th ed. Norwalk, CT: Appleton and Lange, 1992.

This is an excellent, concise textbook of histology, cytology, and microanatomy, with valuable functional explanations. Although it refers mainly to human beings, it is an excellent reference for all mammalian organ systems and tissues.

Monti-Bloch, L., and Grosser, L. *Effect of putative pheromones on the electrical activity of the human vomeronasal organ and olfactory epithelium.* Journal of Steroid Biochemistry and Molecular Biology 39: 573–582, 1991.

Adult human beings appear to have a vomeronasal organ, like all other mammals, except whales and porpoises.

Olds, R. J., and Olds, J. R. *A Color-Atlas of the Rat—Dissection Guide.* New York, Toronto: John Wiley & Sons, 1979.

Although not a dissection guide in the original sense, this booklet contains some good color photographs of various stages of dissection and of organs, as well as a section on the biology of the rat.

Popesko, P., Ratjova, V., and Horak, J. *A Colour Atlas of the Anatomy of Small Laboratory Animals,* Vol. 2, *Rat, Mouse, Hamster.* London: Wolfe Publishing, 1990.

This is a true atlas, with large color-coded illustrations; there is no accompanying text.

Rosenberg, H. I. *How to improve the quality of the environment in the undergraduate dissection laboratory.* American Biology Teacher 54: 171–172, 1992a.

Rosenberg, H. I. *How to reduce the level of formaldehyde in the dissection lab.* The Morphology Teacher, No. 21, 1992b.

Sataloff, R. T. *The human voice.* Scientific American, December: 108–115, 1992.

This article presents a new theory of the mechanism of voice production by the vocal cords.

Schwarz, E. *Classification, origin and distribution of commensal rats.* Bulletin of the World Health Organization: 411–416, 1960.

An account of the types of rats that associate with human beings, their origin, and present distribution.

Thorpe, D. R. *The Rat and Sheep Brain.* New York: National Press Books, 1968.

A color photographic atlas.

Walker, W. F., Jr. and Homberger, D. G. *Vertebrate Dissection,* 8th edition. Philadelphia: Saunders College Publishing, 1992.

Walker, W. F., Jr., and Homberger, D. G. *A Study of the Cat with Reference to Human Beings*, 5th ed. Philadelphia: Saunders College Publishing, 1993.

A dissection guide to the cat, with background information on embryonic development, evolution, and function of mammalian organ systems.

Wells, T. A. G. *The Rat—A Practical Guide.* New York: Dover Publications, 1968.

A standard laboratory manual.

Weijs, W. A. *Morphology of the muscles of mastication in the albino rat, Rattus norvegicus.* Acta Morphologica, Neerlando-Scandinavia 11: 321–340, 1973.

Williams, P. L. et al., Eds. *Gray's Anatomy.* 38th British ed. Edinburgh: Churchill Livingstone, 1995.

Although based on human beings, this encyclopedic work treats all aspects of mammalian anatomy (gross, functional, microscopic, and developmental) and includes many comparative observations.

Wineski, L. E., and English, A. W. *Phenoxyethanol as a nontoxic preservative in the dissection laboratory.* Acta Anatomica 136: 155–158, 1989.

Zeman, W., and Innes, J. R. M. *Craigie's Neuroanatomy of the Rat.* New York: Academic Press, 1963.

A revision and expansion of Craigie's classic study of the rat brain, originally published in 1925. Gross and microscopic anatomy of the nervous system of the rat are described in detail, and what was known of brain function at that time is summarized.

External Anatomy, Skin, and Skeleton

RATS AND MICE ARE WIDELY USED IN ANATOMICAL STUDIES because they are the most available adult mammals to dissect. Like human beings, birds, reptiles, amphibians, and fishes, mammals are animals that have a backbone, or vertebral column. This characteristic places all these animals in the **subphylum Vertebrata,** which is a part of a larger group, the **phylum Chordata.** Other subphyla of chordates includes the peculiar marine sea squirts and lancelets.

Rats, human beings, and other members of the **class Mammalia** resemble birds in maintaining a relatively high body temperature and level of metabolism independently from variations in the external temperature. Both groups are, therefore, described as **homeothermic.** They are also called **endothermic** because they generate heat internally by metabolic processes and control its loss at the body surface. Hair (and the air it traps) and subcutaneous fat thermally insulate the body. The rate of heat loss at the body surface is regulated by varying the rate of blood flow through the skin and by sweating and panting. Female mammals suckle their young with milk secreted by mammary glands.

All mammals resemble each other in more ways than they differ, so the important features that characterize the group can be observed by studying any mammalian species. Rats and mice belong in the **order Rodentia,** the gnawing mammals. Rodents are characterized by a single pair of enlarged incisor teeth at the front of both upper and lower jaws. Human beings, the great apes, monkeys, and lemurs belong in the **order Primates.**

Most laboratory rats are an albino strain of a wild rat, the Norway rat. The Norway rat, *Rattus norvegicus,* and the Black or Roof rat, *Rattus rattus,* are distinct species. Both rat species probably evolved in some part of Eurasia but have accompanied human beings as they migrated throughout the world. Rats reached North America from Europe with early human settlers. The Norway rat is now widespread in North America, but the Black rat is found mainly in port cities and shipyards. Where they occur together in buildings, the Norway rat prefers the basement and lower floors; the Black rat prefers the upper parts of the building. Laboratory mice are an albino strain of the house mouse, *Mus musculus.*

These laboratory exercises refer primarily to the albino rat, but most of the directions and descriptions apply to the mouse as well, unless otherwise noted.

A. EXTERNAL FEATURES

Like that of most other mammals, the body of the rat or mouse is covered by **hairs,** outgrowths of the skin that entrap a still layer of air and help to insulate the body surface against excessive loss of heat. Small mammals, such as rodents, have a large heat-losing surface relative to their heat-producing mass, so heat conservation is of primary importance. Excess heat, such as that generated during physical exercise, can be dissipated, when necessary, by the evaporation of sweat secreted by microscopic sweat glands in the skin or by panting or both. Human beings have large numbers of sweat glands on most parts of the body, but the heavily furred rats and mice have sweat glands only on the naked pads of their feet.

The body consists of four general regions: **head, neck, trunk,** and **tail.** Two pairs of appendages originate from the trunk. Notice that the length of the tail is slightly less than that of the head and trunk combined. This feature distinguishes the Norway rat from the Black rat, which has a tail that is longer than its head and trunk combined. The Black rat is a better climber than the Norway rat and, like other climbing mammals, uses its long and heavier tail as a balancing organ.

External nostrils (*nares*), mouth, eyes, and ears are evident on the head. The **mouth** is bounded by fleshy and hair-covered **lips** (Fig. 1-1A). Fleshy lips and cheeks characterize most mammals and allow the young to suckle by enabling them to establish an airtight seal around the nipple of the mother. The upper lip is cleft by a vertical groove called the **philtrum.** Notice that the upper lips roll into the mouth behind the large upper incisor teeth and form a wall separating the incisor teeth from the rest of the mouth cavity. In this way, a rat can use its incisor teeth to gnaw through wood or soil without getting indigestible material into its mouth. When gnawing, the pair of enlarged lower incisors scrape against the caudal surface of the upper incisors. The incisor teeth grow continuously and wear away as fast as they grow. A sharp cutting surface is always maintained.

The **eyes** are protected by upper and lower **eyelids.** Spread these apart and notice a third eyelid, the **nictitating membrane,** covering the anterior part of the eye. In a live animal, this membrane moves across the surface of the eye and helps to keep the eye clean. Human beings have only a vestige of this membrane, called the semilunar fold, in the medial corner of the eye.

An external ear flap, the **auricle,** or pinna, helps to direct sound waves to the ear canal, the **external auditory meatus.** The tympanic membrane lies out of sight at the base of this meatus. The auricle is sparsely furred and is an area of potential heat loss. This rate of heat loss can be regulated by varying the rate of blood flow through the thin skin of the auricle. In general, the auri-

cle is smaller in mammals that live in cold climates than in members of the same species living in warm climates.

Long "whiskers," technically called **vibrissae,** grow out from the snout, above the upper eyelid, cheek, and chin. Vibrissae are surrounded by touch receptors at their roots and serve as tactile organs. The vibrissae are very sensitive to even slightly tactile stimuli and to air movements. Vibrissae are especially long and prominent in nocturnal mammals, such as rats, which are most active in the dark, and provide important sensory clues for orientation.

The mammalian trunk consists of a cranial part, the **thorax,** which is encased by ribs and houses the heart and lungs, and a caudal part, which is called the **abdomen** ventrally and the **lumbar region** dorsally. Liver, stomach, and intestines lie within the abdomen. The termination of the digestive tract, the **anus,** lies at the caudal end of the abdomen ventral to the base of the tail. If your specimen is a male, you will see a pair of large sacs, the **scrotum,** extending caudally between the hind legs (Fig. 1-1A). They are very noticeable during breeding periods, because the testes migrate into them at this time. They are less conspicuous during nonreproductive periods, when the testes are withdrawn into the body cavity. The hair covering them is very sparse, a characteristic ensuring that the testes are kept cool for normal sperm development (see Exercise 5). The **penis** is surrounded by the foreskin, or **prepuce,** and lies just cranial to the scrotum. A small part of the penis may protrude beyond the body surface. You can detect the main part of the organ by palpating the skin caudal to the **preputial orifice.**

If your specimen is a female (Fig. 1-1B), the **urinary aperture** lies just dorsal to a penislike protuberance, the **clitoris;** and the opening of the vagina lies in a depression, the **vulva,** just caudal to it. Sexes can be difficult to distinguish, because the protruding parts of the penis and clitoris are about the same size. The distance from the clitoris to the anus, however, is only about one-half that from the penis to the anus. Mature females have well-developed nipples (**mammary papillae**) associated with the mammary glands. One group of nipples, usually three pairs, is found on the chest, or thoracic region; another group of about three pairs is found on the caudal abdominal region.

The tail of a rat or a mouse is covered by **horny scales** similar to those that cover the body surface of a reptile. Small hairs emerge in groups between the scales. If you examine your own skin, you will see that hairs are not evenly distributed but tend to emerge in clusters of two or three. Very probably, during the evolution of mammals from a reptilian ancestor, hairs developed between the reptilian scales, the scales became reduced or lost, and the skin was covered by hair.

Terrestrial vertebrates typically have five **digits.** The first digit, or thumb, of a rat or mouse is reduced

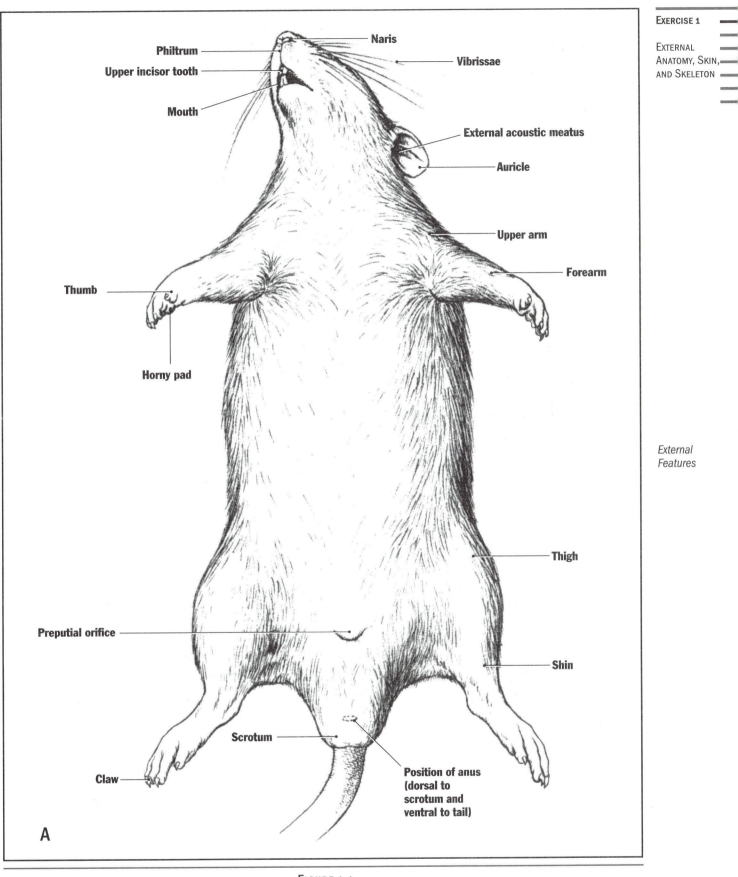

*External
Features*

A

FIGURE 1-1
External anatomy of the rat. (A) Ventral view of a male rat.

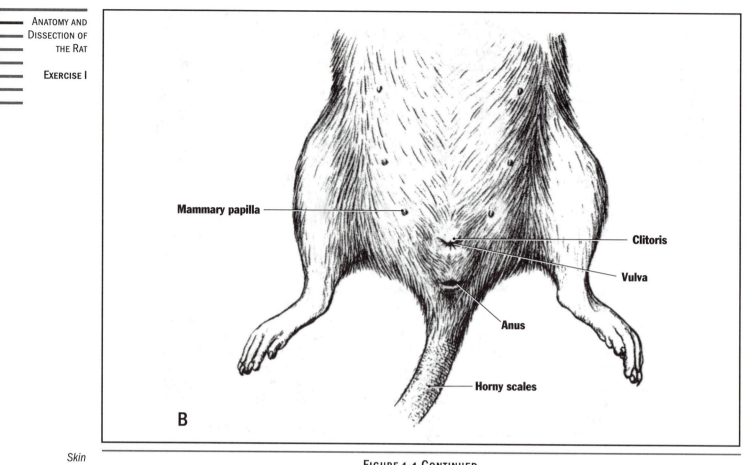

B

FIGURE 1-1 CONTINUED
External anatomy of the rat. (B) Ventral view of the caudal end of a female rat.

and has a small **nail** instead of a **claw,** but the other fingers and all the toes are well developed and clawed. Rodents walk on the horny **digital pads** and **footpads,** which are at the ends and bases of their digits and on the soles of their feet. The skin that covers the footpads and digit pads of rats bears microscopic dermal ridges that are comparable to the dermal ridges that are arranged in whorls on the fingertips and inner surface of the palms in human beings. These dermal ridges increase the friction of the skin, thereby allowing human beings to grasp objects securely and rats to run over smooth surfaces without slipping.

The segments of the forelimb, or **pectoral appendage,** and hindlimb, or **pelvic appendage,** are similar to ours and can be seen and felt: **brachium** (upper arm), **antebrachium** (forearm), and **hand** in the pectoral appendage; **femur** (thigh), **crus** (shin), and **foot** in the pelvic appendage. The foot is much longer than the hand. During locomotion, rats walk on their digits, with the wrist and heel raised off the ground. This posture is called **digitigrade.** Humans walk with the entire foot flat on the ground, a posture called **plantigrade.**

B. SKIN

The skin, or **integument,** is the most exposed boundary layer of the body and forms the interface between the animal's internal environment and the outside world. Because vertebrates maintain an internal environment that is distinctly different from the external environment, the integument is an important organ. It protects the body against abrasion, undue loss of water and salt, ultraviolet radiation, and other assaults by the external world. But the integument does not completely isolate the body from the environment. Many sensory signals are received through the integument, and mammalian integument plays an important role in thermoregulation.

Study the structure of the skin and hair on a histological slide of a vertical section through the skin of a mammal. The structure of the skin correlates closely with its function.

The skin consists of two layers: a thin, outer **epidermis,** and an inner, much thicker **dermis** (Fig. 1-2). By using high power on the microscope, you can see that

4

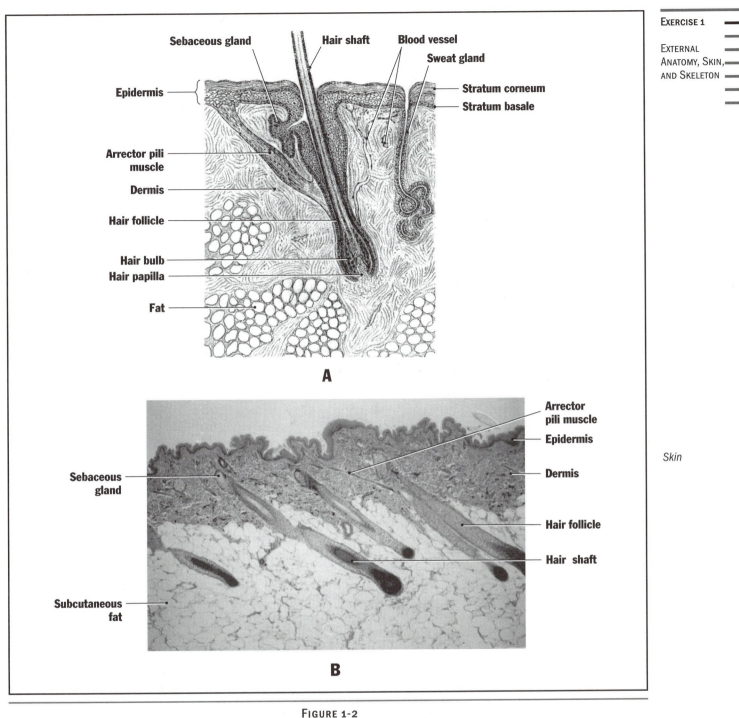

A

B

Skin

FIGURE 1-2
Mammalian skin. (A) Diagram of a vertical section through the skin.
(B) Low-power photomicrograph of a vertical section through the skin.

the epidermis is a stratified squamous epithelium. An epithelium is a tissue in which the cells are tightly packed together. Materials that pass through an epithelium must go through the cells, which have a regulatory role. For this reason, epithelium covers all body surfaces and lines nearly all body cavities. Epithelial cells rest upon an inconspicuous **basal lamina** of minute intercellular fibers. The **stratum basale** of the epidermis, next to the dermis, consists of cuboidal cells. (Cell boundaries are often difficult to see in animal tissues, but the size and orientation of a cell can often be inferred from the size and shape of its more conspicuous nucleus.) Because

5

cells in the stratum basale divide mitotically and produce more cells, this layer is sometimes called the **stratum germinativum.** Dividing cells can sometimes be seen. As many of the new (daughter) cells move toward the surface, they become flattened and impregnated with a horny protein, **keratin,** that binds with lipids and is impermeable to water. Eventually the cells die and form a horny **stratum corneum** at the skin surface that reduces the loss of body water through evaporation. The stratum corneum layer loses cells at the surface as rapidly as new ones are added to it. Intermediate layers can be recognized in thicker regions of the skin. Pigment is produced by specialized cells that lie in the deepest part of the epidermis or penetrate into it from the dermis. It is unlikely that these will be seen, but **pigment granules** are deposited among the epidermis cells and absorb some light energy.

The dermis consists of a dense, fibrous connective tissue. **Connective tissue** is characterized by an extensive extracellular matrix that has been secreted by the cells. The matrix consists of a viscous ground substance containing many fibers. Bundles of intercellular fibers can be seen running in many directions. Most of these fibers are composed of the protein **collagen,** which is relatively inextensible, but some are composed of a different, more extensible and resilient protein called **elastin.** The cellular components of fibrous connective tissue are scattered *Skin* **fibroblasts,** of which only the nuclei can be seen. Bundles of striated muscle fibers, representing integumentary muscles that move the skin, are found in the deeper parts of the dermis in many mammals. Fat tissue, an important thermal insulation, may also be found in this region and in the subcutaneous tissue.

Hair follicles extend into the dermis from the epidermis and sometimes into the subcutaneous tissue. (Because the plane of the section of a slide seldom parallels the axis of the hair, as it does in Fig. 1-2A, most of the hairs will be seen in oblique sections, as shown in Fig. 1-2B.) A hair follicle consists of a stratified epithelium that is supported by connective tissue, and its epithelium is continuous with the epidermis. Each follicle completely surrounds a **hair shaft**—a shaft of cornified epithelial cells that are continually produced in the hair bulb at the base of the follicle. Pigment granules may be seen in the hair shaft and bulb. In some slides, bundles of smooth muscle fibers, the **arrectores pili muscles,** can be seen. They attach to the follicles and extend toward the skin surface to be anchored in the dermis directly under the epidermis. When the body temperature falls, the arrectores pili muscles contract and pull the follicles, which are normally oriented somewhat obliquely to the plane of the skin surface, into a more nearly erect position. This position enables the hairs to entrap more air between them and to form a more effective insulating layer. The arrectores pili muscles can also

be stimulated by fear or other emotions, with the same "hair-raising" effect. In human beings, contraction of the arrectores pili muscles generate "goose bumps" by pulling down the dermis near the bases of the hairs.

Skin structure varies somewhat with body region, so, depending on the region from which your sample came, skin glands, nerves, and blood vessels may be seen. Skin glands develop by growing into the dermis from the epidermis, so their secretory cells are epithelial in origin. **Sebaceous glands** are associated with hair follicles in some regions of the body. If present in your slide, they appear as saclike clumps of lighter-staining cells attached to, or lying beside, the follicles (Fig. 1-2B). Many of their cells are filled with oil droplets, which are discharged into the follicle when the cells break down. This oil conditions hair and keeps it healthy. In human beings, it also prevents excessive drying of the skin surface and keeps the skin soft and smooth.

Sweat glands are long, coiled, tubular glands. They are not common in most of the skin of heavily furred mammals, but the slides may include some. They appear in sections as small clumps of cuboidal epithelial cells (Fig. 1-3A). The nuclei of the cells are relatively large and round. In examining a cross section, you may see that the cells surround a very small cavity, or lumen, but shrinkage of the tissue in the course of slide preparation sometimes obliterates the cavity. Flattened nuclei peripheral to the cuboidal cells belong to elongated **myoepithelial cells,** whose contraction compresses the glandular tubule and assists in the discharge of sweat. Sweat contains some salts and excretory products, but its main component, water, is secreted in abundance when body temperature rises, and its evaporation on the skin surface removes much body heat. Many heavily furred mammals have few sweat glands and reduce body temperature by panting, that is, by evaporating water from the saliva on the tongue surface.

Blood vessels differ in appearance from sweat glands. They have a relatively large lumen that is lined by a single layer of thin, squamous epithelial cells, whose nuclei are much smaller than those of cuboidal cells and are frequently flattened (Fig. 1-3B). The wall of a **capillary** consists only of a single layer of squamous epithelial cells, but that of a small **artery** or **vein** will also contain some smooth muscle and connective tissue. Arteries have thicker walls containing more muscle fibers than the corresponding veins. Therefore, arteries usually have a round cross section in slides, whereas the thinner walled veins are more compressed and often have an irregular cross section. The amount of blood flow through vessels near the surface of the dermis plays an important role in regulating body temperature. Body heat is conserved by a constriction of these vessels and a reduction of blood flow when ambient temperatures are low; opposite changes occur when body temperature is high.

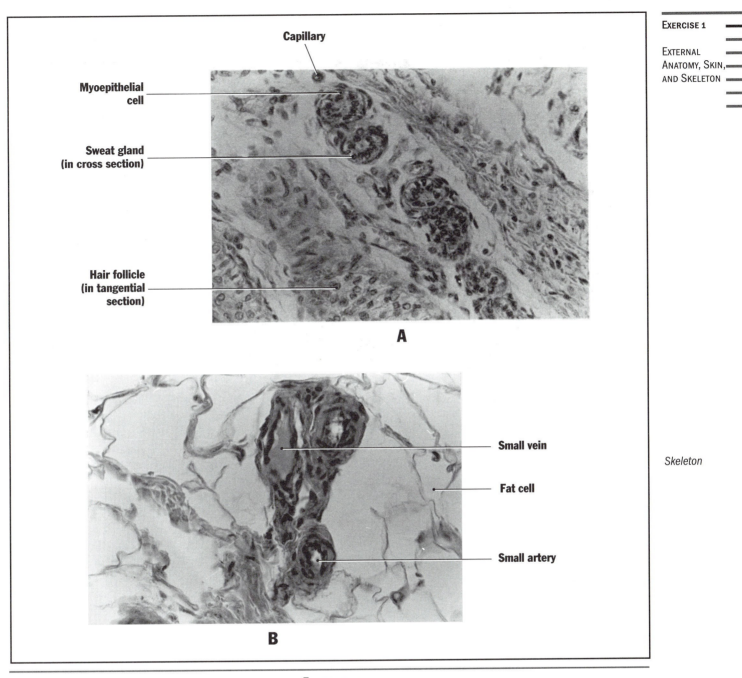

Capillary

Myoepithelial
cell

Sweat gland
(in cross section)

Hair follicle
(in tangential
section)

A

Small vein

Fat cell

Small artery

B

Skeleton

FIGURE 1-3
Mammalian skin. (A) High-power photomicrograph of sections of a
sweat gland. (B) High-power photomicrograph of a cross section through two
small arteries and an accompanying vein surrounded by subcutaneous fat.

C. SKELETON

The skeleton of vertebrates consists of bones and cartilages bound together by ligaments. It forms the supporting framework of the body. Beyond this, many skeletal
parts, such as the skull, protect deeper organs; many
form lever arms that transmit muscle forces to some
point of application, such as the feet, and many house
blood-forming tissue (red bone marrow).

Study and compare the mounted skeletons of a rat
and a human being (see Figs. 1-8 and 1-9). Because both
are mammals, they are similar in their basic features.

Differences in the skull reflect differences in the brain, sense organs, and feeding mechanisms. Differences in the postcranial skeleton reflect differences in support and locomotion, which are a consequence of the type of posture—quadruped or biped.

The skeleton of vertebrates is internal, for it develops within the body wall or in deeper tissues. It is not a superficial secretion on the body surface as is the exoskeleton of invertebrates such as crustaceans and insects.

The part of the skeleton in the body wall and appendages is known as the **somatic skeleton.** It consists of **dermal bones** that develop embryonically directly within connective tissue in or just beneath the dermis of the skin, such as many superficial bones in the skull, and deeper **cartilage replacement bones,** which are formed when bone replaces embryonic cartilaginous skeletal elements. Anatomists divide the somatic skeleton into the **axial skeleton,** in the longitudinal axis of the body (skull, vertebral column, ribs, and sternum), and the **appendicular skeleton** (bones of appendages and the bones of the shoulders and hips, which form the supporting girdles).

Skeleton

Vertebrates also have a **visceral skeleton,** which forms from cartilages in the cranial part of the gut wall (i.e., the pharynx). The visceral skeleton supports the gills and jaws of fishes and participates in the feeding and breathing movements of these animals. It is greatly reduced in mammals, consisting primarily of the hyoid bone, which supports the base of the tongue, of the auditory ossicles, which transmit sound waves from the tympanic membrane across the middle ear cavity to the inner ear, and of the cartilages of the larynx.

C.1. Skull

Examine a skull of a rat and compare it with a human skull (Figs. 1-4, 1-5, and 1-6). A skull consists of a **cranial region** (**cranium**) housing the brain and parts of the ear, and a **facial region** containing the eyes and nose and forming the jaws. Because of our large brain, the cranial region of a human skull is much larger and more globular than that of a rat. The facial region forms a long snout in most mammals but is quite short in human beings. This skeletal change correlates in part with the reduction in our sense of smell, the forward rotation of our eyes, and our bipedal posture.

The large opening at the caudal end of the cranial region is the **foramen magnum,** through which the spinal cord enters the skull to connect with the brain. Because of our upright posture, this foramen has rotated under the human skull. The foramen magnum of all mammals is flanked by a pair of rounded **occipital condyles** by which the skull articulates with the vertebral column. Most of the other small foramina in the skull transmit nerves and blood vessels.

A lateroventral handle of bone, the **zygomatic arch,** forms the lower border of the **orbit,** which contains the eyeball, and a **temporal fossa,** from which the temporal jaw muscle originates. Orbit and temporal fossa merge in the rat but are separated by a plate of bone in human beings.

The large opening just ventral to the caudal end of the zygomatic arch is the **external acoustic meatus.** In life, a tympanic membrane lies at the base of the external acoustic meatus and separates it from the **tympanic cavity,** or middle ear cavity. The tympanic cavity houses the auditory ossicles and, in the rat, is encased ventrally by a bubble of bone called the **tympanic bulla.** In a human skull, only a flat plate or bone lies in the comparable position.

The small opening that you can see on the ventral surface of the rat skull at the rostromedial corner of the bulla is the **auditory canal.** This canal carries the auditory tube, which connects the tympanic cavity with the pharynx and equalizes air pressure on the inside and outside of the tympanic membrane. The inner ear, which contains the receptive cells for detecting sound and changes in body position, lies within a buttress of bone within the cranial cavity, which is known as the **otic capsule.** You can glimpse this capsule by looking into the cranial cavity through the foramen magnum.

The large bump of bone in a human skull located caudal to the external acoustic meatus is the **mastoid process.** Rats have a similar but smaller process. Its large size in humans correlates with the attachment of certain large neck muscles that help support and balance the head atop the neck and shoulders. A pointed **styloid process** extends ventrally from the external acoustic meatus in the human skull. It serves for the attachment of a ligament suspending the hyoid bone. Although rats also possess such a ligament, they do not have a styloid process.

A pair of **external nostrils,** or *nares,* lead from the front of the face into paired **nasal cavities.** The passage from the rostromedial corner of one orbit into a nasal cavity is the **nasolacrimal canal,** a passage carrying a duct through which surplus tear fluid is drained.

Examine the ventral surface of the skull and notice that the nasal cavities are separated from the mouth cavity by the **hard palate** (Fig. 1-6). **Internal nostrils** (*choanae*) lie dorsal to the hard palate and lead into a part of the pharynx. They are obviously paired in the human skull, but this is difficult to see in the rat skull. The hard palate and its fleshy continuation, the soft palate (Exercise 3), separate the air and food passages from each other and allow a mammal to manipulate food within its mouth while still breathing.

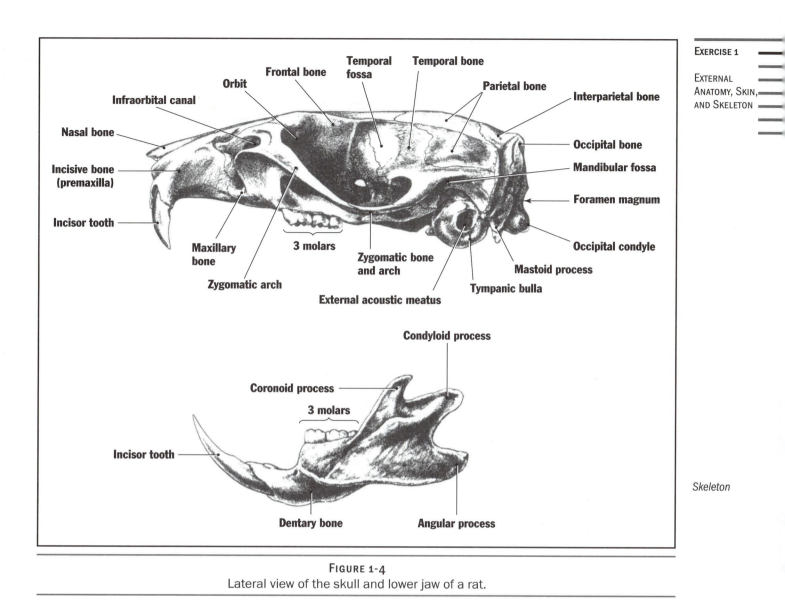

FIGURE 1-4
Lateral view of the skull and lower jaw of a rat.

Skeleton

The rat hard palate is perforated by two large, oval **palatine fissures** (Fig. 1-5). Certain blood vessels and nerves and a small pair of incisive ducts pass through them into the nasal cavities. The incisive ducts lead into paired, tubelike vomeronasal organs in the floor of the nasal cavities (Exercise 7). In mammals in which the vomeronasal organs are well developed, they serve to detect certain oderiferous secretions produced by other members of the same species (pheromones) that play a role in intraspecific regulation of social and reproductive behavior. These organs are reduced in human beings.

The groove at the caudal end of the zygomatic arch is the **mandibular fossa** (Fig. 1-4). The condyle on the condyloid process of the lower jaw articulates here. The mandibular fossa is shallower and flatter in the rat than in human beings. This shape allows the rat's jaw to move more freely fore and aft during gnawing.

Examine the lower jaw, or **mandible.** Its caudal end bears three processes. The middle **condyloid process** carries the smooth condyle that articulates with the mandibular fossa. The temporal jaw muscle, which originates from the temporal fossa, inserts on the small, dorsal **coronoid process.** A masseter muscle arises from the zygomatic arch and inserts on the ventral **angular process** and adjacent parts of the lower jaw. By closing your jaws tightly, you can feel the bulging of the temporal and masseter muscles at the temples and corner of your jaw bone, respectively.

The teeth are deeply set in sockets in the jaw margins. On each side of the upper and lower jaws of an adult human being, there are two **incisors,** one **canine,** two **premolars** (bicuspids), and three **molars**

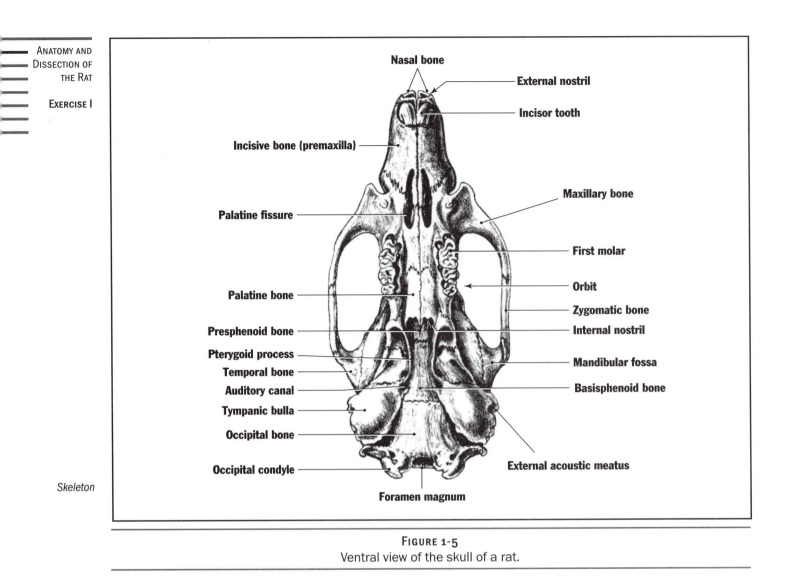

Nasal bone

External nostril

Incisor tooth

Incisive bone (premaxilla)

Maxillary bone

Palatine fissure

First molar

Orbit

Palatine bone

Zygomatic bone

Presphenoid bone

Internal nostril

Pterygoid process

Temporal bone

Mandibular fossa

Auditory canal

Basisphenoid bone

Tympanic bulla

Occipital bone

External acoustic meatus

Occipital condyle

Foramen magnum

Skeleton

FIGURE 1-5
Ventral view of the skull of a rat.

(Fig. 1-6B). The canine of humans has become incisiform, resembling an incisor, and no longer forms a long stabbing tooth as it does in many mammals. Its root, however, is much longer than that of an incisor.

The dentition of rodents is highly specialized for gnawing and grinding. A rat has only one greatly enlarged incisor on each side of each jaw, and there is a wide gap (**diastema**) between the incisor and the next teeth, the three molars (Fig. 1-4). Other teeth have been lost in the course of evolution. Except for the incisor teeth of rodents, the teeth do not grow as the jaw enlarges during postnatal development.

Mammals have two sets of teeth. A **milk set** is replaced during childhood by larger **permanent teeth.** The molar teeth are not replaced but emerge later, in human beings at the approximate ages of 6, 12, and 18. The jaw reaches its full size by the time the last molar erupts.

Many of the individual bones that compose the skull are shown in Figs. 1-4, 1-5, and 1-6.

C.2. Vertebrate, Ribs, and Sternum

A vertebra from the thoracic region of a rat or a human being is illustrative of the fundamental structure of all vertebrae (Fig. 1-7). The ventral part of a vertebra is a solid disk of bone, the **vertebral body,** or centrum. A **vertebral arch** lies dorsal to the vertebral body and encases the spinal cord. Various processes, to most of which muscles and ligaments attach, extend from the arch. A **spinous process** extends dorsally, and a pair of **transverse processes** extend laterally. A pair of **caudal articular processes** extend caudally from the vertebral arch of one vertebra to overlap a pair of **cranial articular processes** extending cranially from the vertebra just behind the one being examined. On the side of the ver-

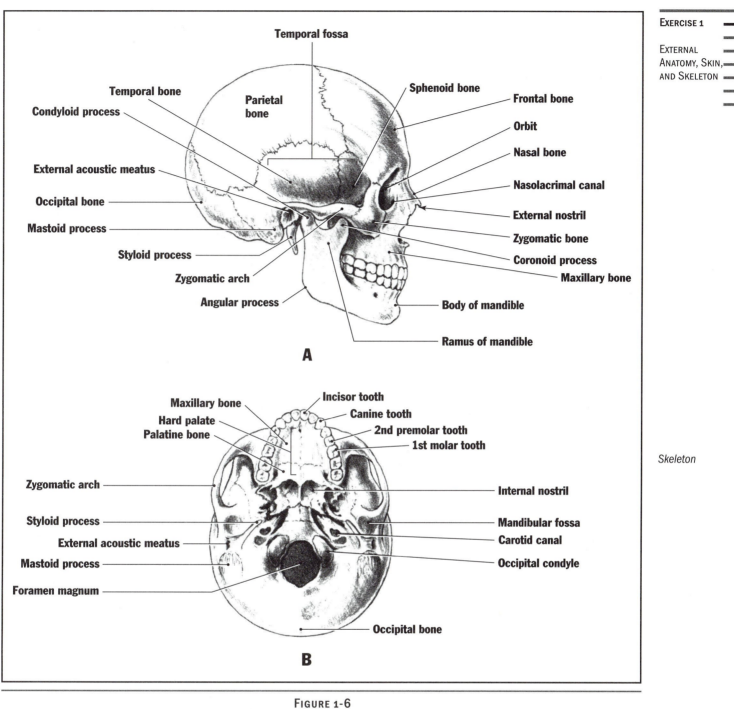

A

B

Skeleton

FIGURE 1-6
Human skull. (A) Lateral view. (B) Inferior view.

tebral column of an articulated skeleton, small holes can be seen between the bases of successive vertebral arches. These are the **intervertebral foramina,** through which the segmental spinal nerves emerge. Cartilaginous **intervertebral disks,** which are filled with a fibrous and gelatinous material, separate successive vertebral bodies.

The vertebral column as a whole is basically a beam that supports the head and body and transfers the weight to the appendages. The vertebrae are modified according to different regional requirements of support and movement. Five vertebral regions can be recognized (Fig. 1-8). Nearly all mammals have a freely movable neck and seven **cervical vertebrae.** The first two cervical

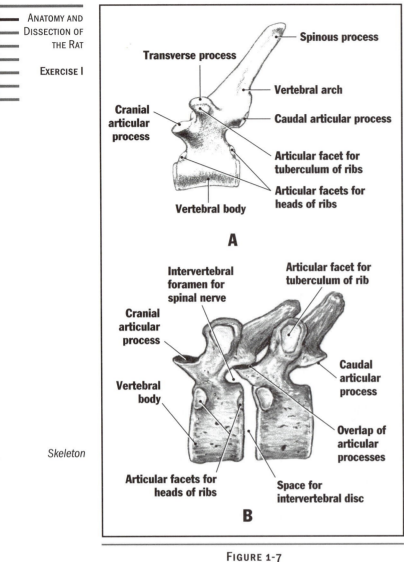

Skeleton

FIGURE 1-7
Lateral views of mammalian thoracic vertebrae;
anterior is toward the left. (A) Individual vertebra
of a rat. (B) Two articulated human vertebrae.

vertebrae, the **atlas** and **axis,** are specially modified to form a sort of universal joint for head movements. The transverse processes of cervical vertebrae include a small embryonic rib that has fused onto them, and most of the processes are perforated by a canal through which a vertebral artery and vein run. Rats and mice have 13 **thoracic vertebrae,** to which the **ribs** are attached. Most ribs have a **tuberculum,** which articulates with the transverse process, and a small **head,** which unites with the vertebral column between successive vertebral bodies so that one part of its articulation is on the caudal end of one vertebral body and the other part is on the cranial end of the next caudal vertebral body (Fig. 1-7B). The ribs terminate distally in flexible **costal cartilages,** most of which unite with the breast bone, or **sternum,** to form a thoracic rib cage that contributes to body support, protects the heart and lungs, and, by rib movements, helps in ventilating the lungs. Six **lumbar vertebrae** are caudal to the thoracic vertebrae and have large transverse processes, with which embryonic ribs have fused. **Sacral vertebrae,** usually three or four in number, are fused into a solid **sacrum** that articulates with the pelvic girdle and transfers weight to the hind legs. Weight is transferred to the front legs by a muscular sling that extends from a point near the distal ends of the ribs to the pectoral girdle. Rats have from 26 to 30 **caudal vertebrae** in the tail, which become progressively incomplete distally until only the vertebral bodies are present. Human beings (Fig. 1-9) have the same types of vertebrae as rats but their number varies slightly: 7 cervical, 12 thoracic, 5 lumbar, 5 sacral, and from 3 to 5 reduced caudal vertebrae that are fused into a small **coccyx** to which certain pelvic muscles attach.

C.3. Pectoral Girdle and Appendage

A flattened, triangular **scapula** is the largest component of the shoulder, or **pectoral girdle** (Fig. 1-8). The **scapular spine,** a sharp ridge to which muscles attach, extends along its lateral surface and terminates near the shoulder joint in a process known as the **acromion.** The scapula has a socket, the **glenoid fossa,** that articulates with the humerus of the upper arm. The small hook of bone dorsal and medial to this cavity is the **coracoid process.** It is the remnant of a larger and independent coracoid bone present in nonmammalian vertebrates. A slender **clavicle,** which helps to brace the scapula, extends from the acromion to the cranial end of the sternum.

The upper arm bone, or **humerus,** of the rat and mouse has a very prominent crest of bone on its cranial surface, the **deltoid tuberosity,** to which the deltoid muscle of the shoulder attaches. Human beings, too, have a deltoid tuberosity, but it is much smaller. Prominent crests of bone, such as this one, increase the mechanical advantage of muscles attaching to them, for they function as lever arms.

The bones of the foreleg are an **ulna** on the lateral side and a **radius** on the medial side. The ulna has a prominent process, the **olecranon,** that extends behind the elbow joint and receives the insertion of the triceps muscle of the upper arm. The front foot is normally in a prone position (the sole facing the ground), but the radius can rotate around the ulna to bring the foot into a supine position (sole facing upward).

Eight small **carpals,** which are difficult to distinguish, constitute the wrist portion of the front foot, five elongated **metacarpals** lie in the region of the sole or palm, and small **phalanges** are in the free part of each digit. Mammals have two phalanges in the first digit (the thumb) and three in each of the others.

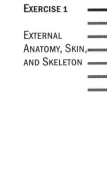

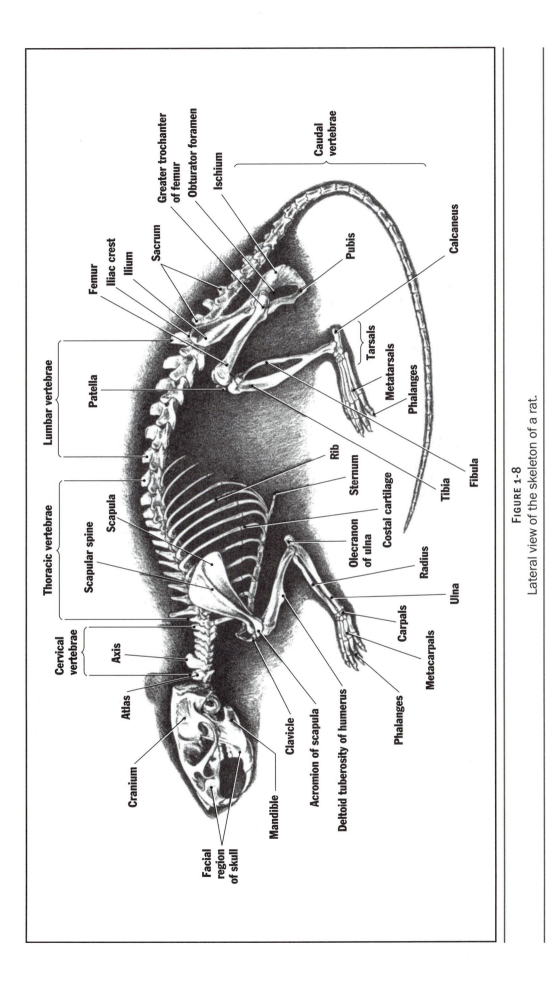

FIGURE 1-8

Lateral view of the skeleton of a rat.

Skeleton

Skeletal
Materials

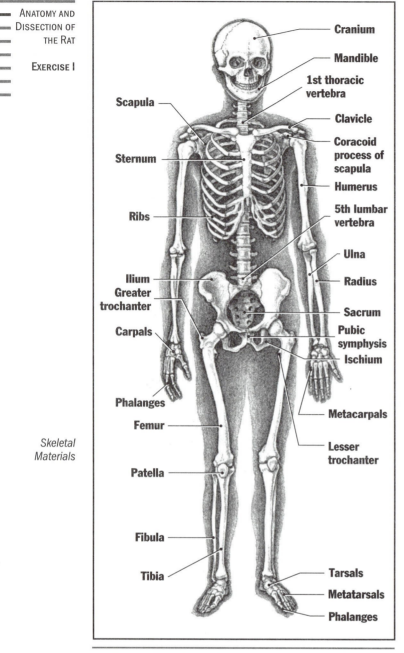

FIGURE 1-9
Anterior view of the human skeleton.

C.4. Pelvic Girdle and Appendage

The hip or **pelvic girdle** (Fig. 1-8) consists of three bones that meet and fuse with each other at the socket (the **acetabulum**) for the hip joint. The long, bladelike **ilium** extends cranially and dorsally from the acetabulum to articulate with the sacrum. The **ischium** extends caudally and ventrally, and the **pubis** extends cranially and ventrally. Together, these bones constitute the hip bone, or **os coxae.** The ilium bears a prominent **iliac crest** on its craniodorsal end to which certain pelvic muscles

attach. The ischium and pubis are separated from each other by a large opening, the **obturator foramen,** which in life is closed by a sheet of connective tissue from which certain pelvic muscles arise. The two pubic bones, one from each side of the body, join ventrally to form the **pubic symphysis.** The pelvic girdle on each side and the sacrum enclose the **pelvic canal,** through which the digestive tract and urinary and reproductive structures pass. The human pelvis contains the same parts, but the proportions of the bones are different because of our upright posture. The ilium, in particular, is relatively shorter but very broad and flaring. This shape provides a large surface from which the buttock (gluteal) muscles, which help stabilize the body atop the legs, originate. The human pelvis also supports the abdominal organs.

The bone of the thigh is the **femur.** Its head articulates with the acetabulum. At the proximal end of the femur are two prominent bony processes, the **greater** and **lesser trochanters,** and a crestlike **third trochanter** extends from the greater trochanter onto the shaft of the femur. Certain pelvic muscles attach onto them.

In most mammals, a large **tibia** lies on the medial side of the shank, and a slender **fibula,** on the lateral side. In rodents, these two bones are fused distally.

Eight small **tarsal bones,** which cannot easily be distinguished in ligamentous skeletons, occupy the ankle part of the foot. The largest of these, the **calcaneus,** extends caudally beyond the ankle joint to form a heel that serves as a lever arm onto which powerful muscles (the gastrocnemius and soleus) from the caudal surface of the shank attach. Five elongated **metatarsals** occupy the region of the sole, and **phalanges** are present in the free parts of the digits.

D. SKELETAL MATERIALS

D.1. Cartilage

Cartilage and bone are modified connective tissues that form the skeleton of vertebrates. Both resist compression well, but bone is more resistant to tension and twisting. Cartilage is more compliant and can grow from within through expansion, without the complex remodeling and replacement required in bone.

Examine a slide of cartilage. It consists of cells, **chondrocytes,** that lie in spaces, the **lacunae,** in an **extracellular matrix** that is secreted by the chondrocytes (Fig. 1-10). The matrix consists of **collagen** fibers, which are masked in a ground substance that contains a great deal of bound water. Cartilage is not supplied by blood vessels and receives nutrients and oxygen by diffusion from its surface. This lack of an internal blood supply limits the size of cartilaginous elements.

Translucent, glasslike cartilage is found, among other places, on the ends of the long bones of the limbs and is known as **hyaline cartilage** (Fig. 1-10A). Visible, elastic fibers consisting of the protein **elastin** are abundant in **elastic cartilage** (Fig. 1-10B), which is found, for example, in the auricle of the ear. The smoothness and compliance of cartilage is a result of its high water content.

Cartilage grows internally by the division of the chondrocytes, which then separate and produce more

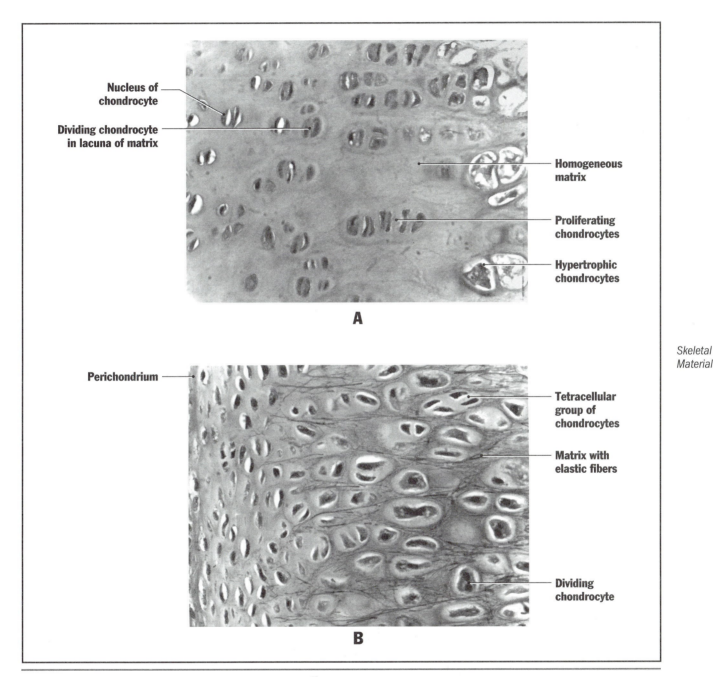

Skeletal Materials

Nucleus of chondrocyte

Dividing chondrocyte in lacuna of matrix

Homogeneous matrix

Proliferating chondrocytes

Hypertrophic chondrocytes

A

Perichondrium

Tetracellular group of chondrocytes

Matrix with elastic fibers

Dividing chondrocyte

B

FIGURE 1-10

Photomicrographs of cartilage (450 ×). During slide preparation, the cartilage cells shrink, leaving a space between them and the surrounding matrix. (A) Hyaline cartilage from a long bone of a kitten. Articular cartilage at the end of the bone is on the left, and this merges with the cartilage of an epiphyses plate on the right. An epiphyses plate is a plate of cartilage located near the end of a long bone of a young mammal where growth in length of the bone occurs. Cartilage cells in the plate proliferate rapidly, and begin to hypertrophy and show other changes as the cartilage is replaced by bone. (B) Elastic cartilage from the external acoustic meatus of a dog.

matrix. It also grows peripherally through the transformation of fibroblasts into chondrocytes in the connective tissue **perichondrium,** which surrounds cartilage. These growth processes allow rapid expansion; therefore, cartilage forms much of the skeleton of embryonic vertebrates. Cartilage persists in adults where a smooth, firm, and compliant tissue is needed, such as on the articular surfaces of the bones and in the costal cartilages.

D.2. Bone

The major skeletal material of adult vertebrates is bone. Bone is a highly vascularized and mineralized, dense connective tissue. Bone is particularly strong because its mineral component resists compression and its fibrous component provides some compliance and resists tension and twisting. Bone, like cartilage and fiberglass, is a composite material made of several substances with different properties. The spread of cracks that may start in a composite material is usually blunted at the interface between two different components.

Living bones are covered by a layer of connective tissue called the **periosteum** (Fig. 1-11). **Compact bone** forms the periphery of the long bones of a limb, and **spongy bones** lies deep to this near the ends of the bone. Marrow fills the large medullary cavities of bone. Bone, like cartilage, consists primarily of an extracellular matrix, but the matrix is composed of hard inorganic salts (mostly crystals of a calcium and phosphorus compound known as **hydroxyapatite**). These crystals are bound to fibrous proteins, chiefly collagen, in the matrix.

*Skeletal
Materials*

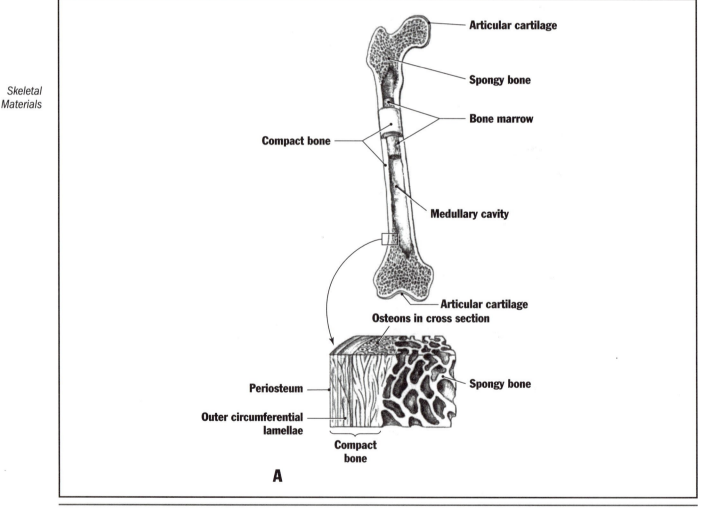

FIGURE 1-11
(A) Structure of the femur, a representative bone.

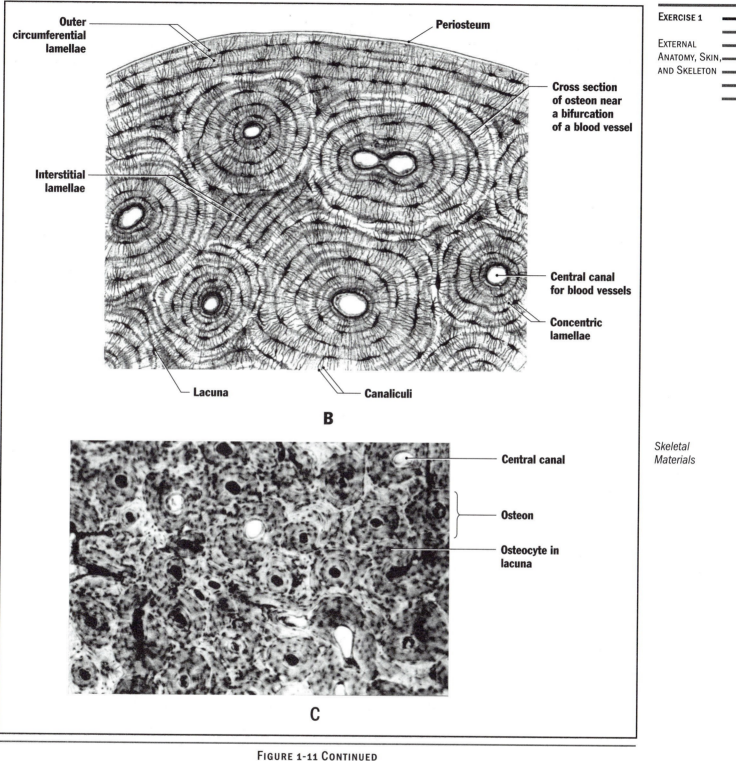

Outer circumferential lamellae

Periosteum

Cross section of osteon near a bifurcation of a blood vessel

Interstitial lamellae

Central canal for blood vessels

Concentric lamellae

Lacuna

Canaliculi

B

Skeletal Materials

Central canal

Osteon

Osteocyte in lacuna

C

FIGURE 1-11 CONTINUED
(B) Enlargement of a group of osteons, as seen in cross section of compact bone.
(C) A photomicrograph of a transverse section through dried compact bone.

Examine a slide of a cross section of compact bone (Figs. 1-11B and C). Most of the bony material is deposited in concentric rings around tiny blood vessels that tend to parallel the longitudinal axis of the bone. Each of these columnar units, which appear circular in cross section, is an **osteon,** or Haversian system. In the center of each is a **central canal** containing, in life, one or more blood vessels (usually capillaries and venules) embedded in loose connective tissue. Around this canal are concentric layers of bone matrix, the **concen-**

tric lamellae. Between the lamellae are rows of dark dots, the **lacunae,** which in life contain the cellular elements of the bone (**osteocytes**). Minute **canaliculi** extend more or less radially from the lacunae and contain processes of the osteocytes. Bone is a dynamic tissue, and much of it is reabsorbed and rebuilt continuously. Parts of former osteons (**interstitial lamellae**) can be seen between the ones formed more recently. Outer **circumferential lamellae** lie at the bone surface.

*Skeletal
Materials*

EXERCISE

Two

Muscles

MUSCLES CONSTITUTE THE LARGEST ORGAN SYSTEM in the body as measured by volume. They are present in all parts of the body and, ultimately, they are responsible for all movements of and within the body, such as the movements of the bony skeleton, the movements of food through the digestive tract, the movement of blood through the circulatory system, and the release of urine from the urinary bladder. Some muscles are inconspicuous because they are part of the skin or of walls of inner organs. Other muscles are very conspicuous and organlike. For example, the heart consists almost completely of a special type of muscle, and the muscles attaching to the skeleton are individually identifiable. We can classify muscles into two or three groups, depending on their position within the body, their innervation, and their histological structure (see Table 2-1 and Section K).

We will dissect only the major skeletal muscles in this exercise. Most of these muscles are **somatic muscles** and are associated with the somatic part of the skeleton. Some muscles of the larynx, jaw, and shoulder are, technically speaking, **visceral muscles.** These are also called **branchiomeric muscles** because they are derived from muscles that were associated with the gill (branchial) apparatus of ancestral fishes.

A. REMOVING THE SKIN

The skin must be removed before we can look at the skeletal muscles. Put your specimen belly-down on your tray and, with a sharp scalpel, make a middorsal incision through the skin at the level of the abdomen. Do not cut too deeply; you will have reached the desired depth when you can grab the freshly cut edge of the skin with a pair of forceps and lift it away from the loose connective tissue that overlies a tight, whitish layer of connective tissue. Now extend your incision at this depth cranially to the base of the head and caudally to the base of the tail. From there, you must extend the incision around the neck at the base of the head, around the tail and genital area, and down the lateral surface of each leg.

Before you remove the skin from the body, notice a thin sheet of cutaneous musculature adhering to the underside of the skin, especially in the abdominal area. Several parts can be distinguished within this layer of cutaneous muscles. The **cutaneus trunci** arises from the midventral line of the trunk and from certain pectoral muscles in the armpit, spreads over the trunk, and inserts on the skin and to the base of the tail. This muscle moves and tightens the skin over the trunk. It is present in most mammals, but human beings lack this muscle.

Over the lateral and ventral surfaces of the neck, the cutaneous musculature continues as the **platysma.** As it spreads over the head, it breaks up into numerous **facial muscles,** which move the lips, nose, eyelids, and the ears. The platysma and the facial muscles are branchiomeric muscles, in contrast to the cutaneus trunci, which is a somatic muscle. The facial muscles are very well developed in human beings and are responsible for facial expressions. You will see the facial muscles later when removing the skin from the head. As you carefully remove the skin, cut the blood vessels and nerves that supply it. Notice that those blood vessels and nerves along the trunk emerge from the body at regular intervals. This pattern is one indication of the basic segmentation of the vertebrate body, a feature less obvious in mammals than in fishes. You will also have to transect

the cutaneus trunci at the base of the tail and as it passes into the armpit.

If your specimen is a mature or lactating female, very extensive **mammary glands** are found adhering to the underside of the cutaneous muscle; ducts lead from the glandular tissue to the nipples by passing through the layer of cutaneous muscles. The lobulated mammary tissue is divided into distinct cranial and caudal portions. The abdominal portion supplies the abdominal nipples and extends caudally between the hindlegs to the anal region. The thoracic portion of the mammary gland supplies the nipples on the chest and extends cranially between the forelegs, behind the elbows, and on the sides of the neck up to the base of the ears!

The connective tissue that surrounds the cutaneous muscles and the mammary glands and attaches them to the skin also connects them to the skeletal muscles. It may also contain fat tissue, especially in well-fed specimens, and is known as the **superficial fascia.** Remove as much of it as you can from the side of the body on which the muscles are to be studied. A large mass of **brown fat** lies in the middle of the back between the two shoulder blades. This specialized fat is a center for nonshivering heat production. It is activated by a severe lowering of body temperature, and warm blood flowing from it goes quite quickly to vital thoracic organs. Brown fat is particularly important to the newborn, to all rodents during cold adaptation, and to certain rodents during arousal from hibernation. You may remove it. A tough, more highly organized **deep fascia** encases some muscles, especially in the lumbosacral middorsal region; it should be left in place for the time being.

B. MUSCLE TERMINOLOGY AND MUSCLE FUNCTION

An individual skeletal muscle consists of many muscle fibers. Each muscle fiber comprises a multitude of myofibrils, which contain the contractile elements of muscles

TABLE 2-1
Classification of Muscles

GENERAL LOCATION	INNERVATION	HISTOLOGICAL TYPE	SPECIFIC LOCATION	RELATIONSHIP TO GENERAL BODY PLAN
Skeleton	Voluntary	Striated	Most skeletal muscles	Somatic
			Some larynx, jaw, and shoulder muscles	Branchiomeric (special visceral)
Inner organs and hair follicles	Involuntary	Smooth	Most inner organs and hair follicles	Visceral (general visceral)
		Cardiac	Heart	

(see Section K). Muscle fibers are held together by connective tissue to form **muscle fiber bundles** (fascicles), which also are held together by connective tissue to form muscle bellies (Fig. 2-1). The smallest units discernible with the naked eye are muscle fiber bundles. Near the attachment of a muscle to a bone, each muscle fiber attaches to collagen fibers through a **myotendinous junction.** The collagen fibers are held together by connective tissue to form a cordlike **tendon** for a spindle-shaped muscle, or a sheetlike **aponeurosis** for a sheetlike muscle. The tendon fibers and their surrounding connective tissue attach to a bone by interweaving with the connective tissue of the periosteum enveloping the bone and by sending collagen fibers into the bone substance for anchoring.

Although a contracting muscle exerts an equal force on both skeletal elements to which it attaches, by convention, its attachment to the more stationary and proximal skeletal element is called its **origin,** and its **insertion** is the site where the muscle is attached to the more mobile and distal skeletal element. An individual muscle may have more than one origin and insertion. It

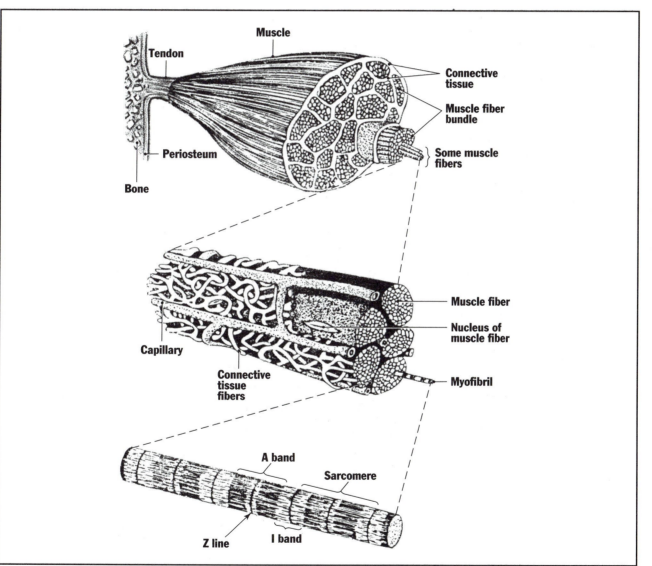

Muscle Terminology and Muscle Function

FIGURE 2-1

Diagram of the structure of a skeletal muscle, its attachments to bones, and its microscopic composition. (Redrawn from W. F. Walker, Jr. and Homberger, D. G. *Vertebrate Dissection,* 8th ed. Philadelphia: Saunders College Publishing, 1992. After Dorit, R., Walker, W. F., Jr. and Barnes, R. D. *Zoology.* Philadelphia: Saunders College Publishing, 1991. And Fawcett, D. W. *Bloom and Fawcett, A Textbook of Histology,* 12th ed. New York: Chapman & Hall, 1994.)

is not always easy to determine which attachments of a particular muscle are its origin and insertion. In such cases, it is better to use the neutral term of **attachment** for all ends of a muscle.

Because muscles perform work only by contracting, there must be an antagonistic mechanism that restores the contracting muscle to its original length. Usually the antagonist is another muscle or muscle group. One muscle group moves a skeletal element in one direction; the **antagonistic muscle** group moves it in the opposite direction. The following terms define the more common antagonistic actions:

1. **Flexion** and **extension.** Flexion is a bending; that is, it is the moving of a distal part toward a more proximal part. Extension is the opposite action (Fig. 2-2).
2. **Protraction** and **retraction.** Protraction is a forward movement of the entire limb at the shoulder or hip. Although the term *flexion* is sometimes used for this movement, protraction is more accurate for a quadruped. Retraction is a backward movement of the entire limb.
3. **Adduction** and **abduction.** Adduction is the pulling of a part toward a point of reference. Abduction is the movement away from the point of reference. The

midventral line of the body is the point of reference for adduction or abduction of the limbs.

Many students are at first overwhelmed by the many unfamiliar names for muscles, but these names can be very helpful because they are generally descriptive of the attachments (sternomastoid), shape (trapezius), number of divisions (triceps), or function (levator scapulae) of muscles.

C. DISSECTION TECHNIQUE FOR MUSCLES

Only the more conspicuous muscles can be closely observed in an exercise of this scope. Before attempting to identify certain muscles, carefully remove fat and connective tissue from their surfaces. Then identify the borders of individual muscles by carefully watching the orientation of the muscle fiber bundles, because it often changes in adjacent muscles. Once you have identified the borders of various muscles, you may separate the muscles from one another by using two pairs of forceps to break the connective tissue fibers that bind the mus-

Dissection Technique for Muscles

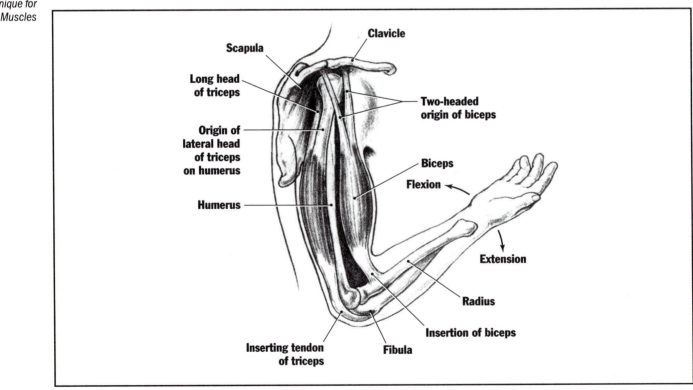

FIGURE 2-2
Diagram of an antagonistic pair of muscles
in the human arm and their actions.

cles together. If you proceed correctly with your dissections, the surfaces of the muscles will remain smooth and intact.

When it is necessary to transect a superficial muscle to reveal a deeper one, you may first want to ascertain the muscle fiber direction of the superficial muscle and free its borders from connective tissue. Lift one border with a pair of forceps and cut with a pair of scissors (keep the blunt tip on the lower half of the scissors) through the center of the muscle as far as you can. If necessary, free the muscle from the underlying muscles before you extend the cut through the entire muscle at a right angle to its muscle fiber direction. Bisect only one muscle at a time. Turn bank (reflect) the two ends. This procedure will facilitate the reconstruction of the superficial muscles when you wish to review them.

D. MUSCLES OF THE SHOULDER

The first muscles to be considered are those that extend from the trunk to the pectoral girdle, and from the trunk and pectoral girdle to the proximal end of the humerus. Collectively, they move the shoulder and the arm as a whole; or, when the front foot remains in a fixed position on the ground, they move the trunk relative to the foot.

D.1. Superficial Muscles of the Shoulder
A group of three trapezius muscles covers the lateral surface of the shoulder (Fig. 2-3). The line of origin of the muscles extends from the back of the skull caudally along the tops of the spinous processes of the cervical vertebrae and most of the thoracic vertebrae. Different parts of the complex insert on different parts of the shoulder: the cranial **cleidocervicalis** (clavotrapezius) goes to the clavicle; the middle **cervical trapezius** (acromiotrapezius) goes to the acromion on the scapula; and the caudal **thoracic trapezius** (spinotrapezius) goes to the upper part of the scapular spine. These muscles help to hold the shoulder in place and to rotate the pectoral girdle forward and backward. They form a single muscle, the trapezius, in human beings. Most muscles of the shoulder are somatic muscles, but this group is branchiomeric.

A ribbon-shaped **omotransversarius** (levator scapulae ventralis) also inserts on the acromion. As it extends forward, it goes deep to the cleidocervicalis and arises from the base of the skull and from the atlas. It, too, pulls the scapula forward. Human beings do not have this muscle.

The deltoid complex is a triangular muscle mass lying ventral to the trapezius group. In rats, but not in human beings, it consists of two parts: a **spinodeltoid,** which originates from the scapular spine, and a **cleido-**

brachialis (clavodeltoid), which originates from the clavicle. Both converge to insert on the deltoid tuberosity of the humerus. The more cranial cleidobrachialis protracts the humerus, the spinodeltoid retracts it, and together they abduct it.

The large, triangular muscle that lies caudal to the shoulder and arm and fans out over the back is the **latissimus dorsi** (Fig. 2-3). It arises deep to the thoracic trapezius from the spinous processes of thoracic vertebrae and from part of the deep fascia, the **thoracolumbar fascia,** which covers the dorsal trunk muscles. It passes into the axilla to insert on the proximal end of the humerus. It is an important retractor of the arm.

The pectoral group consists of two muscles that arise from the sternum and insert on the deltoid tuberosity of the humerus (Fig. 2-4). In rodents, the first one, the **pectoralis superficialis,** is smaller than the deeper **pectoralis profundus.** (In humans, the superficial one, called the pectoralis major, is larger than the deeper pectoralis minor.) Notice that the pectoralis profundus inserts more proximally on the humerus. The cutaneus trunci joins the pectorals at their insertion. Primary functions of the group are adduction and retraction of the arm.

D.2. Deep Muscles of the Shoulder
Cut through the centers of the trapezius muscles and the latissimus dorsi; reflect their ends. A **teres major** (Fig. 2-5), which arises from the caudodorsal angle of the scapula, lies cranial to the latissimus dorsi and inserts with it on the proximal end of the humerus. The teres major helps retract the arm. A very small teres minor, which you need not try to find, lies deep beside the insertion of the teres major.

The muscle arising from the lateral surface of the scapula cranial and dorsal to the scapular spine is the **supraspinatus.** Its dorsal end is partly covered by the rhomboideus capitis (see the next paragraph). An **infraspinatus,** which is partly covered by the spinodeltoid, arises from the scapula, ventral and caudal to its spine. Both of these muscles insert on the proximal end of the humerus. The former helps to protract the humerus; the latter rotates it outward.

Several deep muscles extend from the dorsal border of the scapula to the trunk and neck. All act upon the scapula, which they help to hold in place and pull cranially and caudally. A ribbonlike **rhomboideus capitis** lies deep to the cranial two parts of the trapezius group and extends from the back of the skull to the craniodorsal angle of the scapula. A thicker **rhomboideus cervicis** (minor) and **rhomboideus thoracis** (major) extend from the spinous processes of cervical and thoracic vertebrae, respectively, to the dorsal border of the scapula. The rhomboideus cervicis and rhomboideus thoracis are difficult to separate from each other, and you need not try to do so. Human beings lack the rhomboideus capitis.

*Muscles of
the Shoulder*

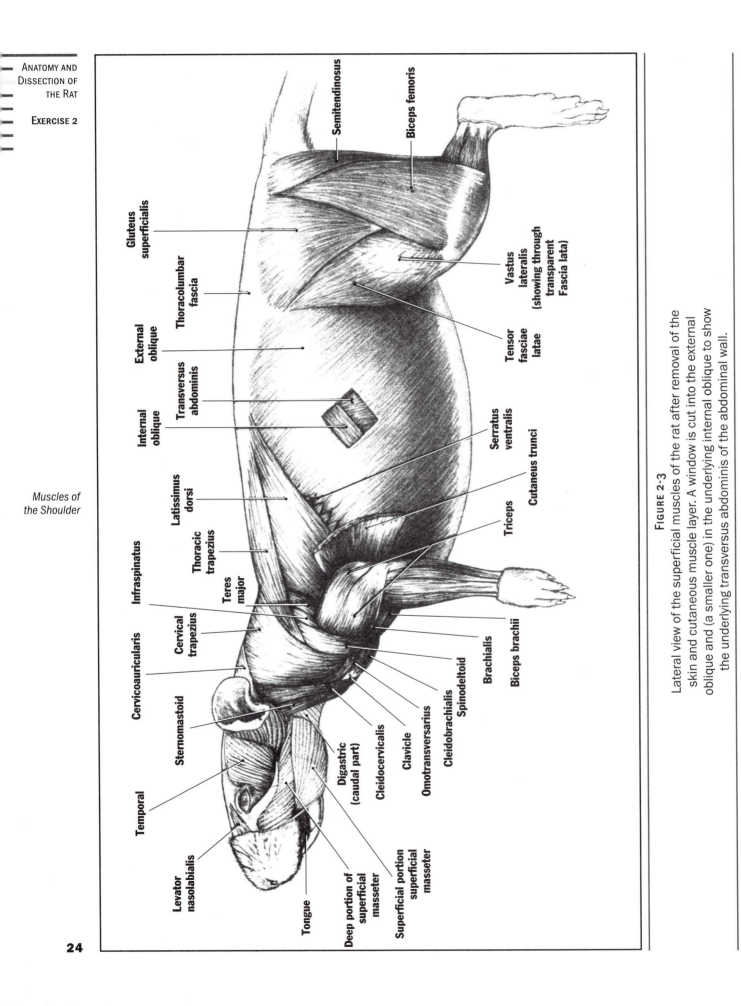

FIGURE 2-3

Lateral view of the superficial muscles of the rat after removal of the
skin and cutaneous muscle layer. A window is cut into the external
oblique and (a smaller one) in the underlying internal oblique to show
the underlying transversus abdominis of the abdominal wall.

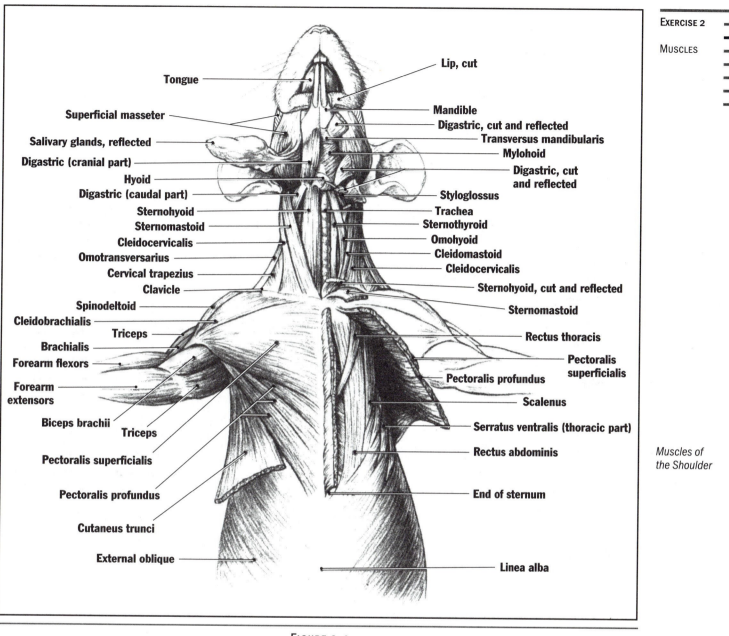

FIGURE 2-4
Ventral view of the muscles of the thorax, shoulder, neck, and head. The superficial muscles are shown on the right side of the animal, the deeper muscles on the left side.

Labels (left side, top to bottom): Tongue, Superficial masseter, Salivary glands, reflected, Digastric (cranial part), Hyoid, Digastric (caudal part), Sternohyoid, Sternomastoid, Cleidocervicalis, Omotransversarius, Cervical trapezius, Clavicle, Spinodeltoid, Cleidobrachialis, Triceps, Brachialis, Forearm flexors, Forearm extensors, Biceps brachii, Triceps, Pectoralis superficialis, Pectoralis profundus, Cutaneus trunci, External oblique

Labels (right side, top to bottom): Lip, cut, Mandible, Digastric, cut and reflected, Transversus mandibularis, Mylohoid, Digastric, cut and reflected, Styloglossus, Trachea, Sternothyroid, Omohyoid, Cleidomastoid, Cleidocervicalis, Sternohyoid, cut and reflected, Sternomastoid, Rectus thoracis, Pectoralis superficialis, Pectoralis profundus, Scalenus, Serratus ventralis (thoracic part), Rectus abdominis, End of sternum, Linea alba

Muscles of the Shoulder

Cut through the rhomboideus complex so that the top of the scapula can be pulled away from the trunk. The **serratus ventralis** is the large fan-shaped muscle that arises from the transverse processes of the cervical vertebrae and, by a series of slips, from the ribs. It inserts on the dorsal border of the scapula deep to the insertion of the rhomboideus cervicis and rhomboideus thoracis. The serratus ventralis has a cervical part and a thoracic part; the cervical part is called the levator scapulae in human anatomy. Besides helping to rotate the scapula, the serratus ventralis forms a sling that suspends the trunk between the shoulder girdles and forelimbs. Except for the clavicle, which connects the sternum to the acromion of the scapula, there is no bony connection between the trunk skeleton and the pectoral girdle.

Cut through the pectoral muscles and pull the arm forward. The muscle that lies between the scapula and serratus ventralis is the **subscapularis** (Fig. 2-6). It arises from the medial surface of the scapula and inserts on the proximal end of the humerus; its primary action is arm adduction.

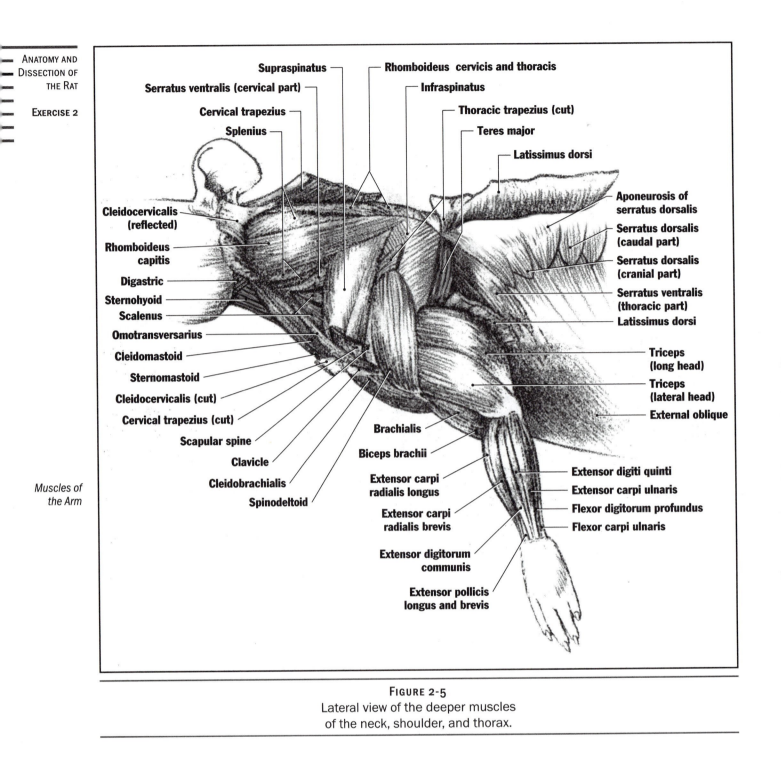

Muscles of
the Arm

FIGURE 2-5
Lateral view of the deeper muscles
of the neck, shoulder, and thorax.

E. MUSCLES OF THE ARM

Now you will examine a group of muscles that cover the humerus. All arise from the pectoral girdle or humerus, and most act upon the forearm. The largest muscle is the **triceps** (Figs. 2-5 and 2-6), which covers most of the caudal, lateral, and medial surfaces of the humerus. The medial surface of the triceps is covered by a thin **tensor fasciae antebrachii,** which arises from the edge of the latissimus dorsi and inserts on the ante-

brachial fascia and on the insertion of the triceps. Bisect it if it has not already been destroyed. The triceps arises by three heads: a long head from the caudal border of the scapula, a lateral head from the proximal part of the humerus, and a small medium head (visible only if you look on the medial side of the arm) from the proximal two-thirds of the humerus. All heads converge to form a powerful tendon that inserts on the olecranon of the ulna. This muscle is the primary extensor of the forearm.

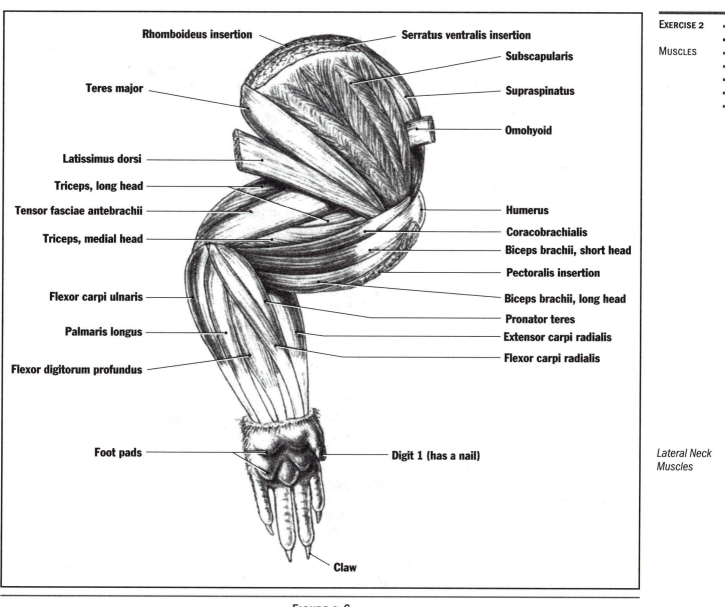

Rhomboideus insertion

Serratus ventralis insertion

Subscapularis

Teres major

Supraspinatus

Omohyoid

Latissimus dorsi

Triceps, long head

Tensor fasciae antebrachii

Humerus

Triceps, medial head

Coracobrachialis

Biceps brachii, short head

Pectoralis insertion

Flexor carpi ulnaris

Biceps brachii, long head

Palmaris longus

Pronator teres

Extensor carpi radialis

Flexor digitorum profundus

Flexor carpi radialis

Foot pads

Digit 1 (has a nail)

Claw

Lateral Neck Muscles

FIGURE 2-6
Medial view of the muscles of the left shoulder and pectoral appendage.

Two smaller muscles cover the cranioventral surface of the humerus: laterally a **brachialis** and medially a **biceps brachii.** The brachialis arises from the proximal part of the humerus and inserts on the ulna. The biceps brachii (Figs. 2-5 and 2-6) arises by two heads: a long head from the edge of the glenoid fossa on the scapula, and a smaller head from the coracoid process of the scapula. The coracoid head can be seen crossing the insertion of the subscapularis. Both heads unite and insert on the radius. The biceps brachii and the brachialis together flex the forearm.

A small **coracobrachialis** (Fig. 2-6) has a common origin with the coracoid head of the biceps brachii but diverges to insert on the humerus. It is a small humeral adductor.

Forearm muscles are not described here, but many can be identified by referring to Figs. 2-5 and 2-6.

F. LATERAL NECK MUSCLES

Muscles on the side of the neck (Figs. 2-3 and 2-4) extend from the pectoral girdle and sternum to the skull. They differ from some of the shoulder muscles in that their primary function is to flex and turn the head rather than to act on the pectoral girdle.

Locate the cleidocervicalis again. The straplike muscle that parallels its cranioventral border is the **sternomastoid.** It originates from the sternum and inserts

27

behind the ear on the mastoid region of the skull. Bisect both the cleidocervicalis and sternomastoid, and reflect them. A **cleidomastoid** extends between the clavicle and the mastoid region deep to the preceding muscles. Another straplike muscle, the **omohyoid,** lies deep to the cleidomastoid and extends between the cranial border of the scapula (Figs. 2-4 and 2-6) and the hyoid. It helps certain hypobranchial muscles (see the next section) pull the hyoid caudally. The sternomastoid, cleidomastoid, and omohyoid are branchiomeric muscles like the trapezius muscle group.

G. THROAT AND LINGUAL MUSCLES

Several ribbonlike muscles extend from the sternum to the hyoid or larynx on the ventral side of the neck (Fig. 2-4). Those to be described pull the hyoid and larynx caudally during swallowing. Origin and insertion of these throat muscles can be inferred from their names. The most superficial muscle is the **sternohyoid.** It is a paired muscle, but its left and right halves are so close together that it appears to be unpaired. Bisect the sternohyoid and the omohyoid, and reflect their ends. The long and very narrow strip of muscle on the lateral surface of the windpipe, or trachea, is the **sternothyroid.** (The thyroid is one of the cartilages in the larynx.) A short **thyrohyoid** extends from the insertion of the sternothyroid to the hyoid.

The more cranial lingual muscles, which will not be studied in this exercise, lie deep between the hyoid and the chin and extend into the tongue. They pull the hyoid cranially and move the tongue.

H. HEAD AND JAW MUSCLES

Carefully remove the skin from one side of the head. **Facial muscles,** which insert onto and move the skin, will be removed with it. Some of these muscles that are associated with the nose and ear—for example, the **levator nasolabialis** and the **cervicoauricularis**—are quite conspicuous (Fig. 2-3). All facial muscles are branchiomeric. So are the remaining muscles of the head that insert and act on the lower jaw. A **temporal** muscle fills the temporal fossa behind the orbit (Fig. 2-3), where it arises from the side of the cranium. It passes deep to the zygomatic arch to insert on the coronoid process of the mandible. It closes the jaw.

Rodents have a particularly large and complex **masseter,** which not only assists the temporal in closing the jaw but also moves the lower jaw back and forth during gnawing and grinding. The **superficial masseter** consists of superficial and deep portions (Fig. 2-3).

The **superficial portion** of the superficial masseter arises from the rostral part of the maxilla and inserts on the lateral side of the mandible as far caudally as the angular process. It is covered by a sheetlike tendon rostrally, and its muscle fiber bundles are oriented more or less horizontally.

The **deep portion** of the superficial masseter arises from the zygomatic arch and inserts on the lateral surface of the mandible next to the insertion of the superficial portion. Its muscle fiber bundles are oriented obliquely from rostrodorsal to caudoventral. Bisect both portions of the superficial masseter, and you will see the **deep masseter** with its almost vertically oriented muscle fiber bundles. Most of the deep masseter arises from the zygomatic arch and inserts on the lateral surface of the mandible dorsal to the insertion of the superficial masseter. A smaller rostral part inserts from the side of the nasal area of the skull and passes through the enlarged infraorbital canal to join the insertion of the rest of the deep masseter. Other jaw-closing muscles arise from the base of the skull and are thus too deep to be seen easily.

A **digastric** muscle (Figs. 2-3 and 2-4) extends from the base of the skull just behind the ear to the chin. It is divided by a central tendon into distinct cranial and caudal parts, called bellies. This is the major jaw-opening muscle.

Detach the chin attachment of the digastric, and carefully dissect beneath it. Two thin sheets of muscle will be seen (Fig. 2-4): A **transversus mandibularis** extends transversely between the two halves of the lower jaw, and muscle fibers of the **mylohyoid** incline slightly caudally to insert on a connective tissue septum in the midline and on the hyoid. These two muscles affect the movement of the two halves of the lower jaw, which are not firmly united at the chin in rodents. The mylohyoid also supports the throat and lingual muscles during movements of the hyoid and during swallowing. Human beings do not have a transversus mandibularis.

I. TRUNK MUSCLES

Trunk muscles can be divided into an **epaxial group,** which lies dorsal and lateral to the vertebral column, and acts to brace and move the back, and a **hypaxial group,** which mainly tightens and moves the walls of the trunk ventral to the vertebral column. Hypaxial muscles will be emphasized in this section.

Three thin sheets of hypaxial muscles form the abdominal wall. They arise dorsally, primarily from the ribs and thoracolumbar fascia, and insert ventrally by aponeuroses on a midventral connective tissue septum, the **linea alba** (Fig. 2-4). They can be distinguished by their positions in relation to one another and by their muscle fiber direction (Fig. 2-3). Fibers of the superficial layer, the **external oblique,** extend from craniodorsal to caudoventral. With the help of a scalpel and a pair of forceps, carefully cut a window through the very thin external oblique in the center of the lateral side of the abdomen. You need to cut only deep enough to see the underlying muscle fibers of the **internal oblique,** which extend from caudodorsal to cranioventral nearly at right angles to the superficial external oblique. Spread apart the muscle fiber bundles of the internal oblique with a pair of fine forceps and notice that the muscle fibers of the underlying **transversus abdominis** are oriented nearly transversely to the long axis of the trunk. All three muscle layers help to support and compress the inner organs in the abdominal cavity, especially during expiration.

A longitudinal band of muscle, the **rectus abdominis,** lies deep to the internal oblique muscle and just lateral to the linea alba. It extends from the pelvic girdle to the cranial ribs, and you can observe it most easily in the thoracic region by reflecting the bisected pectoral muscles on one side (Fig. 2-4).

All the thoracic hypaxial muscles have a role during respiration, for they pull the ribs forward to enlarge the thoracic cavity during inspiration and backward to compress the cavity during expiration. The major inspiratory muscle, however, is the diaphragm, which you will not be able to see until the body cavity has been opened (Exercise 3). A short **rectus thoracis** (transversus costarum) crosses the insertion of the rectus abdominis (Fig. 2-4). Several slips of the **scalenus,** which are difficult to separate from one another, lie dorsal and lateral to the rectus thoracis. They originate on the cervical vertebrae, cross the insertion of the serratus ventralis, and attach to the ribs.

Pull the vertebral border of the scapula laterally and ventrally so that you can examine the chest wall medial to the serratus ventralis. The muscle fibers that extend from one rib caudally and ventrally to the next caudal rib constitute an **external intercostal.** Carefully cut through one of the external intercostals and you will find an **internal intercostal,** with fibers lying at right angles to those of the external intercostal. External intercostal muscles contract during inspiration, internal intercostal muscles during expiration. The external and internal intercostal muscles in the thorax correspond to the external and internal oblique layers in the abdominal wall.

There is only a trace of the transverse muscle layer in the thoracic wall: The **transversus thoracis** lies on the inner surface of the ribs near the midventral line and will be seen when the thorax is opened.

The thin sheet of muscle arising from the thoracolumbar fascia and inserting on the dorsal parts of the ribs, which it covers, is the **serratus dorsalis** (Fig. 2-5). Its cranial portion has a role during inspiration, its caudal portion during expiration.

If you bisect the serratus dorsalis and thoracolumbar fascia, you will see the epaxial group of trunk muscles. Most form powerful longitudinal bands that constitute the **erector spinae** muscle group. A triangular **splenius** (Fig. 2-5) lies deep to the rhomboideus capitis and helps to raise and turn the head.

J. MUSCLES OF THE PELVIS AND THIGH

Unlike the muscles in the pectoral region, which are arranged so that all but a few muscles act across a single joint, many of the pelvic muscles extend across both hip and knee joints and, therefore, can move the thigh and the leg at the same time. When the hind foot is held on the ground, they move the trunk relative to the foot.

J.1. Lateral Muscles of the Hip and Thigh

Four superficial muscles lie on the lateral surface of the hip and thigh (Fig. 2-3). A **tensor fasciae latae** arises from the cranial end of the ilium and fans out to insert on the white sheet of connective tissue, the **fascia lata,** covering the thigh. Because part of this fascia in turn attaches to the tibia, the tensor fasciae latae helps to extend the leg as well as to protract the thigh at the hip joint. A large, fan-shaped **gluteus superficialis** (Fig. 2-7) arises by a sheet of fascia from the ilium and sacrum and converges to insert on the third trochanter of the femur. This muscle is an abductor of the thigh and helps to protract the thigh (if the leg is in a retracted position) or to retract it (if the leg is protracted). In human beings, this is the very large buttock muscle, the gluteus maximus, which helps to hold the trunk erect. The **biceps femoris** and the **semitendinosus** lie caudal to the gluteus superficialis and arise, partly in common, from the caudal end of the ischium and from the adjacent caudal vertebrae. The biceps femoris technically has two points of origin, but the heads are difficult to distinguish in the rat. The biceps femoris expands to form a broad muscle that inserts by an aponeurosis along most of the length of the tibia. It forms the lateral wall of the **popliteal fossa,** the depression behind the knee joint. The band-shaped semitendinosus passes medially to form the medial wall of the popliteal fossa and inserts on the shaft of the tibia. Both the biceps femoris and the semitendinosus flex the shank and retract the thigh.

Bisect and reflect the tensor fasciae latae, gluteus superficialis, and biceps femoris (Fig. 2-7). The **gluteus medius,** which is the largest of the gluteal muscles in the rat, lies deep to the gluteus superficialis. It arises from the lateral surface of the ilium and from the sacrum, and inserts on the greater trochanter of the femur. It can be divided into cranial and caudal parts in the rat. A dis-

Muscles of the Pelvis and Thigh

29

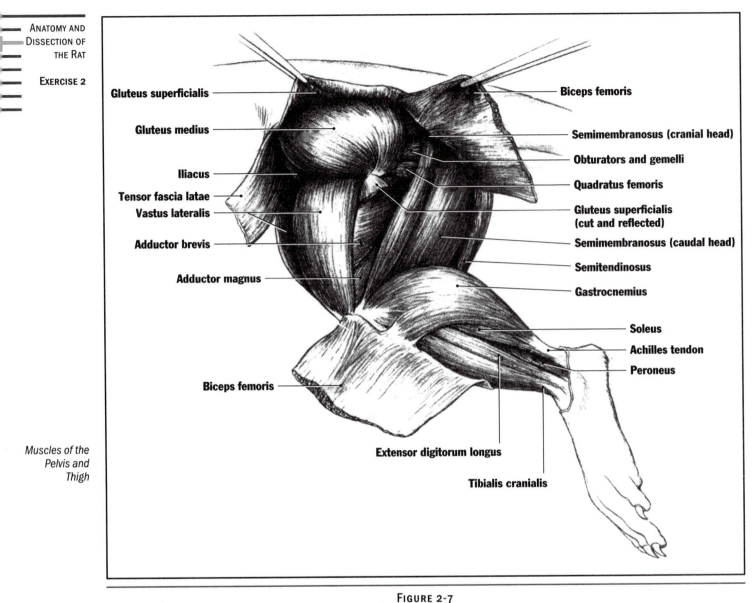

Gluteus superficialis

Gluteus medius

Iliacus

Tensor fascia latae

Vastus lateralis

Adductor brevis

Adductor magnus

Biceps femoris

Extensor digitorum longus

Tibialis cranialis

Biceps femoris

Semimembranosus (cranial head)

Obturators and gemelli

Quadratus femoris

Gluteus superficialis
(cut and reflected)

Semimembranosus (caudal head)

Semitendinosus

Gastrocnemius

Soleus

Achilles tendon

Peroneus

*Muscles of the
Pelvis and
Thigh*

FIGURE 2-7
Lateral view of the deeper pelvic and leg muscles.

tinct piriformis lies deep to the caudal part of the gluteus medius in most mammals, but it is almost inseparably bound with the gluteus medius in the rat. Bisect the cranioventral part of the gluteus medius to locate the **gluteus profundus** (gluteus minimus of human beings). It is a small muscle that arises from the ventral border of the ilium and inserts on the greater trochanter of the femur. Because the gluteus medius and gluteus profundus attach to the greater trochanter dorsal to the hip joint, they are able to retract the leg and to assist the gluteus superficialis in thigh abduction.

Two long, straplike muscles can be found deep to the biceps femoris and cranial to the semitendinosus (Fig. 2-7). The larger and more caudal muscle is the **caudal head** of the **semimembranosus.** It arises from the caudal

end of the ischium and inserts on the proximal part of the medial side of the tibia. The narrower, more cranial muscle is the **cranial head** of the **semimembranosus,** which is very distinct in the rat and can be completely separated from the caudal head. It extends from some of the sacral and caudal vertebrae to the distal end of the femur. Both parts of the semimembranosus assist in thigh retraction, and the caudal head also helps to flex the shank.

Several small and very deep muscles, the **obturators** and **gemelli,** arise from the periphery of the obturator foramen of the pelvis and from the membrane covering this foramen. All insert on the proximal end of the femur between the greater and lesser trochanters. They are difficult to separate in an animal as small as the rat, but part of the group as a whole can be seen between the

30

gluteus group and the semimembranosus (Fig. 2-7). Their primary action is thigh rotation. A slightly larger **quadratus femoris** lies ventral to them and extends between the caudoventral part of the ischium and the lesser trochanter. It helps to rotate and retract the thigh.

J.2. Medial Thigh Muscles

Many of the muscles on the medial surface of the thigh are adductors and pull the leg toward the midventral line of the body. The most superficial of them is the gracilis (Fig. 2-8). In the rat and other rodents, it consists of a **gracilis cranialis,** which originates on the pubis, and a **gracilis caudalis,** which originates on the caudal part of the

ischium. Both extend distally to insert on the proximal end of the tibia; the insertion of the gracilis caudalis is deep to that of the gracilis cranialis.

A triangular **adductor longus** arises from the pubis, crosses the origin of the gracilis cranialis, and inserts on the proximal part of the femoral shaft. If you bisect and reflect the gracilis cranialis, you will see the other muscles of the adductor complex lying cranial to the caudal head of the semimembranosus. An **adductor brevis** arises from the pubis caudal to the adductor longus. It extends deeply into the thigh to insert along the femoral shaft from the third trochanter distally. Its insertion is most easily observed from the lateral side (Fig. 2-7). An **adductor magnus** (Fig. 2-8) arises from the pubis, between the ori-

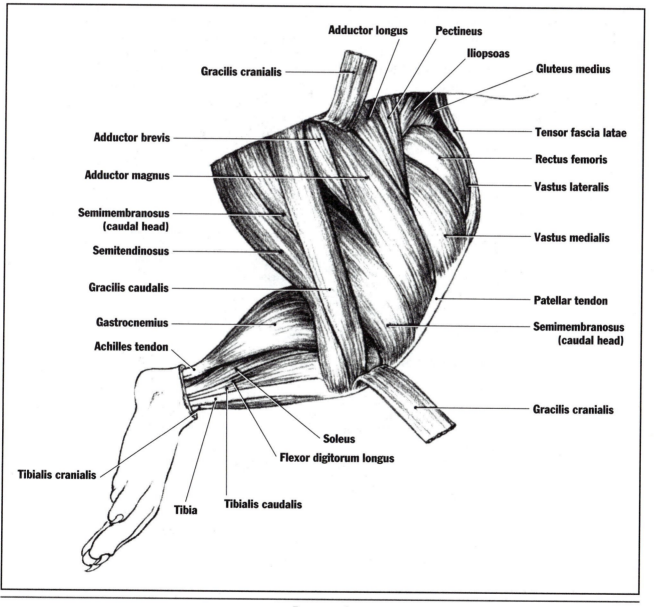

Muscles of the Pelvis and Thigh

FIGURE 2-8
Medial view of the pelvic and leg muscles.

gins of the other two adductors. It is the longest of the adductor group and extends distally to its insertion on the distal end of the femur. A part of this muscle, too, is visible in a deep lateral dissection. A final adductor is the narrow, triangular **pectineus,** which lies cranial to the adductor longus and extends between the front of the pubis and the proximal part of the femoral shaft.

A major protractor of the thigh is the **iliopsoas,** a powerful triangular muscle lying cranial to the pectineus. It arises from the ventral surfaces of the caudal lumbar vertebrae and iliac crest and inserts on the lesser trochanter of the femur.

J.3. Cranial Thigh Muscles

The cranial surface of the thigh is covered by a large **quadriceps femoris.** It consists of four major parts, all of which converge to form the large **patellar tendon** which crosses the knee joint and inserts on the proximal end of the tibia. This tendon slides easily across the joint because it contains a small sesamoid bone, the knee cap, or **patella.** The quadriceps femoris is the major extensor of the leg. One of its heads, the **vastus lateralis,** can be observed best from the lateral side of the thigh (Fig. 2-7), because it arises from the greater and third trochanters of the femur. On the medial side, locate another head, the **vastus medialis** (Fig. 2-8), arising from the proximal part of the femoral shaft just cranial to the insertion of the pectineus. Between these two vasti muscles, on the cranial surface of

Muscle Tissue

the thigh, lies a head called the **rectus femoris.** Because it crosses the hip joint to arise from the ilium, it also acts as a femoral protractor. The fourth head, the small and inconspicuous **vastus intermedius,** arises from the lateral side of the femoral shaft deep to the vastus lateralis. You must bisect and reflect the vastus lateralis to see it. Muscles of the lower leg are not described, but many can be identified from Figs. 2-7 and 2-8.

K. MUSCLE TISSUE

K.1. Skeletal Muscle

Examine a histological slide preparation of a longitudinal section through **skeletal muscle.** The individual muscle cells are cylindrical and very long, with a uniform diameter ranging from 10 to 100 micrometers. Because of their great length, these muscle cells are called **muscle fibers.** In a few cases the muscle fibers may be as long as the muscle they form, but usually they are only a few centimeters long and may be connected in series by connective tissue to extend over the entire length of a muscle. Because these muscle fibers develop embryonically by an end-to-end fusion of embryonic muscle cell precursors (**myoblasts**), which each have one nucleus, each muscle fiber contains many nuclei. The nuclei lie just inside the cell membrane, which is called **sarcolemma** in muscle fibers (Fig. 2-9A,B). The cytoplasm of muscle

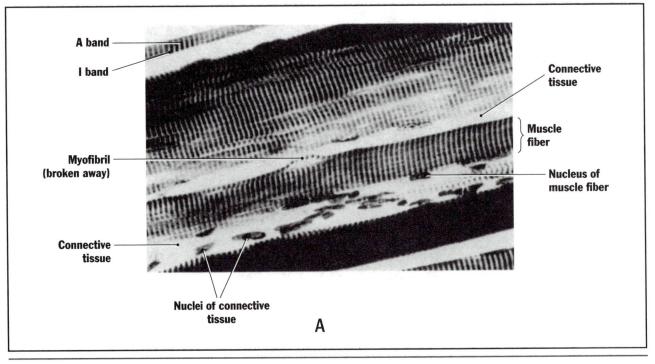

FIGURE 2-9
Skeletal muscle. (A) Photomicrograph at high magnification
of a longitudinal section through skeletal muscle fibers.

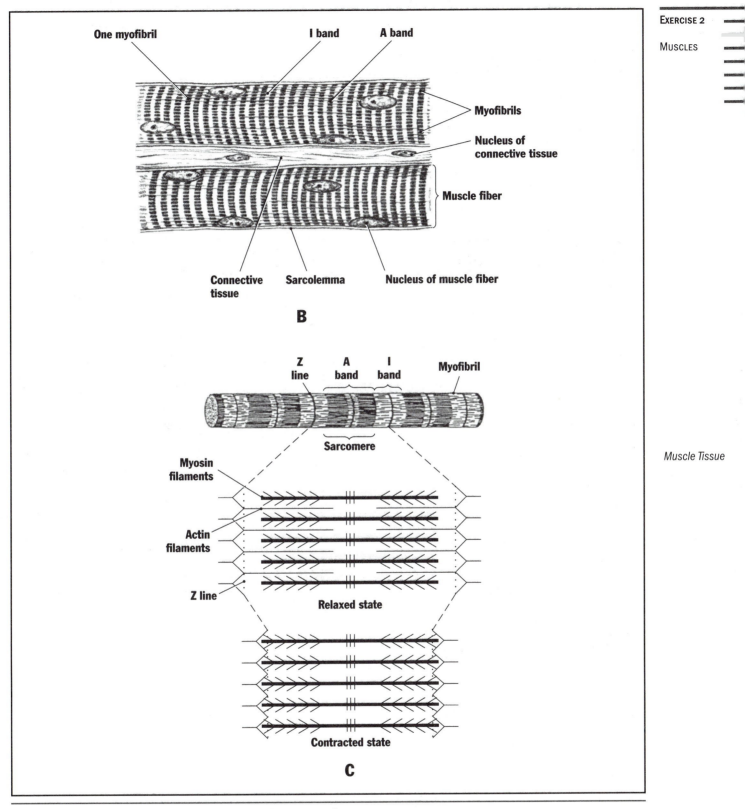

Muscle Tissue

B

C

FIGURE 2-9 CONTINUED

Skeletal muscle. (B) Diagrammatic enlargement of a portion of two skeletal muscle fibers. (C) Diagrammatic view of a myofibril at the microscopic and electronmicroscopic level. (C is redrawn from W. F. Walker, Jr. and Homberger, D. G. *Vertebrate Dissection,* 8th ed. Philadelphia: Saunders College Publishing, 1992. After Dorit, R., Walker, W. F., Jr., and Barnes, R. D. *Zoology.* Philadelphia: Saunders College Publishing, 1991. And Fawcett, D. W. *Bloom and Fawcett, A Textbook of Histology,* 12th ed. New York: Chapman & Hall, 1994.)

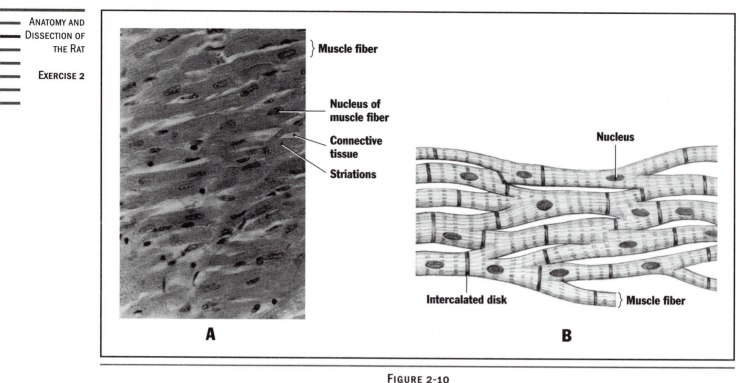

Muscle fiber

Nucleus of muscle fiber

Connective tissue

Striations

Nucleus

Intercalated disk

Muscle fiber

A

B

FIGURE 2-10
Cardiac muscle. (A) Photomicrograph at high magnification of a longitudinal section through cardiac muscle, in which the intercalated disks are not visible. (B) Diagrammatic enlargement of cardiac muscle fibers with intercalated disks.

Muscle Tissue

fibers, called **sacroplasm,** consists almost solely of **myofibrils,** which are the contractile elements of the muscle. They are barely visible with light microscopy because they are only 1 or 2 micrometers in diameter (Fig. 2-9A,B). The alternating dark and light bands that occur along a myofibril are known as **A** and **I bands,** respectively, because of their anisotropic and isotropic properties in polarized light. The A and I bands of the various myofibrils in a particular muscle fiber are in register with one another, so the entire muscle fiber has a banded, or striated, appearance. Because of this appearance, skeletal muscles are often called **striated muscles.**

Electron microscopic and biochemical studies have shown that the myofibrils, in turn, are composed of many ultramicroscopic **myofilaments** of two types, subdivided into contractile units called **sarcomeres** (Fig. 2-9C). The thicker **myosin filaments** are limited to the dark A bands within a sarcomere. The thinner **actin filaments** occupy the light I bands. Each I band is divided by a **Z line** that serves as an anchoring place for the actin filaments of two adjacent sarcomeres. The actin filaments extend into the A bands where they interdigitate with the myosin filaments. When a muscle fiber contracts, complex biochemical interactions between the two types of myofilaments cause the actin filaments to slide deeper into the array of myosin filaments. Consequently, the I bands become narrower, and the myofibrils—and with them the muscle fiber—generate tension and tend to shorten. The precise geometry of the myofilaments is responsible for the great speed and force with which striated muscles contract.

Skeletal muscles are innervated by myelinated nerves that travel within the connective tissue permeating them. Each nerve fiber divides into several terminal branches, each of which attaches to the sarcolemma of one muscle fiber through a **motor end plate** (or myoneural junction). All the muscle fibers that are innervated by a single nerve form a **motor unit.** When a skeletal muscle contracts, only about one-third of all motor units contract at a given time.

K.2. Cardiac Muscle

Examine a histological slide preparation of a longitudinal section through the **cardiac muscle** of the heart wall (Fig. 2-10A). This muscle type, too, is striated, but its striations are much less obvious than those in skeletal muscle. The nuclei are situated in the center of the cells. The individual muscle cells have a diameter of about 15 micrometers. With a length of about 85 to 100 micrometers, they are not

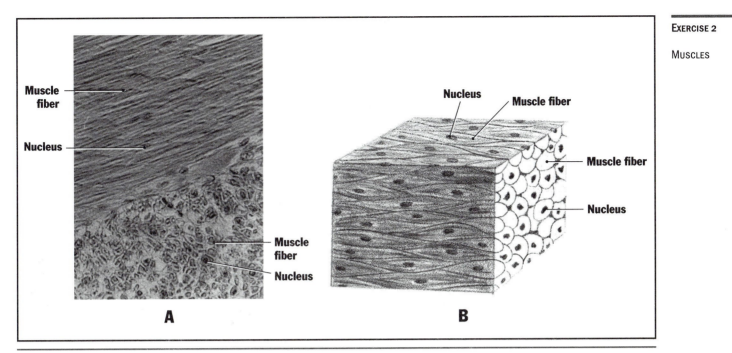

FIGURE 2-11

Smooth muscle. (A) Photomicrograph at high magnification of a longitudinal (upper half) and transverse section (lower half) through smooth muscle. (B) Diagrammatic enlargement of smooth muscle fibers. (B is redrawn from Junqueira, L. C., Carneiro, J., and Kelley, R. O. *Basic Histology,* 7th ed. Norwalk, CT: Appleton and Lange, 1992.)

Muscle Tissue

as long as the muscle fibers of skeletal muscles. The muscle cells branch and are joined to other muscle cells at specialized junctions called **intercalated disks** (Fig. 2-10B).

The individual muscle cells are not innervated by nerves. Action potentials initiated by the pacemaker of the heart (see Exercise 4) spread rapidly through the muscle cells themselves, crossing from cell to cell at the intercalated disks. As a consequence, the cardiac muscle tends to contract as a unit.

K.3. Smooth Muscle

Examine a histological slide preparation of a section through **smooth muscle** (Fig. 2-11A). Smooth muscle is found in the walls of most visceral organs and blood vessels. Smooth muscle fibers are elongated, spindle-shaped cells that are from 20 to 500 micrometers long and have a single nucleus in the center (Fig. 2-11B). The muscle fibers usually are tightly packed and staggered, so the tapered ends of some muscle fibers overlap the thicker center of other muscle fibers. As a consequence, the diameter of smooth muscle fibers varies considerably when the tissue is viewed in cross section. Each muscle fiber contains myofibrils, but they are barely visible in light microscopy. The light and dark bands of the myofibrils within a particular muscle cell are not in register with one another; therefore, the muscle cells do not appear striated.

Nerve fibers terminate on only some of the muscle fibers, so any action potential spreads from the innervated muscle cells to the rest of the muscle cells. However, it does so more slowly than in cardiac muscle. Smooth muscle is well suited for the slow but sustained contractions needed to move food through the digestive tract and to control peripheral blood flow and pressure.

EXERCISE

THREE

Digestive and Respiratory Systems

O RGANISMS ACQUIRE RAW MATERIALS and release some waste products through the digestive and respiratory systems. Specifically, the respiratory system acquires oxygen and releases carbon dioxide (CO_2), a waste product from cell respiration. The digestive system acquires nutrients, such as food, vitamins, and minerals, and releases feces consisting of undigestible materials. Bile pigments, products from the breakdown of senescent red blood cells and the hemoglobin they contain, are also voided with the feces. The nitrogenous wastes of cellular metabolism are removed by the excretory system, which will be studied in Exercise 5.

The digestive and respiratory systems are usually studied together because their parts are often connected and found adjacent to one another. This anatomical proximity is a result of the embryonic development of the lungs as an outgrowth from the pharyngeal part of the digestive tract.

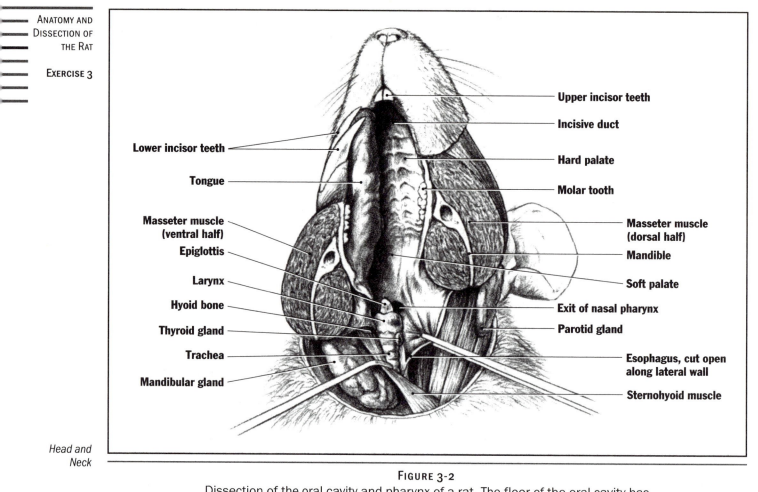

Lower incisor teeth

Tongue

Masseter muscle
(ventral half)
Epiglottis

Larynx

Hyoid bone

Thyroid gland

Trachea

Mandibular gland

Upper incisor teeth

Incisive duct

Hard palate

Molar tooth

Masseter muscle
(dorsal half)

Mandible

Soft palate

Exit of nasal pharynx

Parotid gland

Esophagus, cut open
along lateral wall

Sternohyoid muscle

FIGURE 3-2

Dissection of the oral cavity and pharynx of a rat. The floor of the oral cavity has been turned toward the left side of the drawing and is seen in a lateral view.

which is subdivided into three parts. The **oral pharynx** is the caudal extension of the oral cavity and lies ventral to the soft palate. The **nasal pharynx** is the caudal extension of the nasal cavity and lies dorsal to the soft palate. Both the nasal pharynx and the oral pharynx lead into the **laryngeal pharynx** caudal to the soft palate and above the voice box, or **larynx.** The larynx lies caudal to the tongue and represents the entrance to the windpipe, or **trachea** (see later).

In the laryngeal pharynx, the food and air passages cross. Air passes through the nasal cavities, continues through the nasal pharynx and laryngeal pharynx, and enters the trachea through the larynx. Food and water pass through the oral cavity, continue through the oral pharynx and laryngeal pharynx, and enter the **esophagus,** which lies dorsal to the trachea (see later).

Insert the blunt end of a pair of scissors into the exit of the nasal pharynx and make a longitudinal incision through the soft palate all the way to the hard palate. The nasal pharynx is a single chamber; it is formed by the confluence of the paired nasal cavities. **40** The **internal nostrils** (*choanae*) are the openings

between the nasal cavities and the nasal pharynx; they are seen more clearly on a skull. Near the middle of each lateral wall of the nasal pharynx is the slitlike opening of the **auditory tube** (Eustachian tube) (Fig. 7-4, Exercise 7), which leads to the middle ear cavity.

The middle ear cavity is filled with air. It is necessary to keep the air pressure in the middle ear cavity equal to that of the ambient air so that the delicate tympanic membrane is protected against tension caused by unequal pressures on it and is able to respond to sound waves. The auditory tube allows air to move between the middle ear cavity and the nasal pharynx and, thereby, to adjust the air pressure within the middle ear cavity to that of the ambient air.

The auditory tube and the middle ear cavity develop embryonically from the first pharyngeal pouch. This pharyngeal pouch has evolved from the first gill slit of ancestral fishes.

The larynx is supported by several cartilages, and it has a trough-shaped **epiglottis** at its rostral end. Air enters the larynx easily when its **glottis,** a longitudinal slitlike opening in the center of the larynx is open during inspiration. The windpipe, or **trachea,** extends caudally from the larynx. It is held patent by oval cartilaginous rings in its walls; thus air can easily move through it during respiration. During swallowing, food is prevented from entering the larynx and trachea by the closing of the glottis, an action that causes the epiglottis to fold back on the glottis. Food is pushed into the usually collapsed **esophagus** by a backward movement of the tongue and a rhythmic contraction of the muscular walls of the pharynx. You can follow this swallowing process by putting your fingers on your own larynx, or Adam's apple, while swallowing.

A small mass of glandular tissue, the **thyroid gland,** lies at the cranial end of the trachea. It consists of two lobes, one on each side of the trachea, which are connected ventrally by a narrow band of tissue, the isthmus of the thyroid gland. If your specimen has been injected, the thyroid gland will probably appear pink because it contains many small arteries.

The thyroid gland secretes the hormone thyroxin, which serves to maintain the high rate of metabolism and heat production that is characteristic of mammals. Microscopic **parathyroid glands** are embedded within the tissue of the thyroid gland and secrete a hormone that regulates the calcium metabolism of the body.

Insert a pair of fine scissors into the entrance to the larynx and cut caudally through its dorsal wall and into the trachea. Spread the two halves of the larynx slightly apart and observe the small pair of folds of whitish tissue on the inside of the lateral walls of the larynx. These folds, which extend from the base of the epiglottis dorsally, are the vocal cords. Laryngeal muscles move the vocal cords closer together and apart, and control their tension. The mechanism of vocalization is not based on a stringlike vibration of the vocal cords, as was believed earlier. The vocal cords actually create sound by opening and closing the glottis and, thereby, creating puffs of air during exhalation, in a manner similar to that by which the clapping of hands creates sound.

B. OPENING THE BODY CAVITY

If your animal has not been skinned, first make a longitudinal incision through the skin, slightly to the right of the midventral line, extending from the base of the neck to the caudal end of the abdomen (Incision 1, Fig. 3-3). Use a sharp scalpel for this incision. At this point, be sure to leave the muscular body wall intact. If your specimen is a male, cut to the right of the orifice of the penis, or preputial orifice, when you reach the caudal abdominal region, and stop at the entrance to the scrotum. If your specimen is a female, extend the incision to the right of the urogenital orifice with the clitoris and vulva and stop

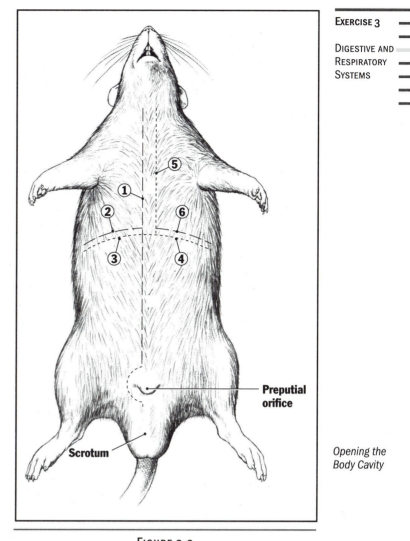

Opening the Body Cavity

FIGURE 3-3
Ventral view of a male rat, showing the incisions that should be made to open the chest and abdomen.

before you reach the anus. Using a pair of scissors, next cut through the clavicle at the base of the neck, which you can palpate after spreading the edges of the cut skin. Insert the blunt end of the pair of scissors into the opening and cut caudally along the skin incision through the ribs and the wall of the thorax and through the musculature of the abdominal wall. Pull up on the body wall as you do this, to avoid injuring the internal organs. Part the incision and find a transverse muscular partition, the **diaphragm,** which separates the chest, or thoracic cavity, from the abdominal cavity.

With a pair of scissors, make a lateral cut through the right half of the body wall just cranially to, and along the attachment of, the diaphragm all the way to the muscles of the back (Incision 2). While doing so, you will cut through the caudal ribs of the thorax. Next make a cut through the right side of the body wall just caudally to, and along the attachment of, the diaphragm (Incision 3). **41**

Extend this incision through the left side of the body wall just caudally to, and along the attachment of, the diaphragm (Incision 4). Now make a longitudinal cut through the thorax slightly to the left of the midline (Incision 5). As you spread apart the edges of Incisions 1 and 5, you will discover a longitudinal membrane (the mediastinal septum, see later), which is attached to the strip of the thoracic wall that was left between the parallel longitudinal incisions; do not damage it. Finally, make an incision through the body wall just cranially to, and along the attachment of, the diaphragm on the left side of the specimen (Incision 6), meanwhile leaving intact the central strip of thoracic wall between Incisions 1 and 5. Spread open the flaps of the thoracic wall and palpate the ribs on the inside of the thoracic wall. With a pair of scissors, cut through each rib near its attachment to the vertebral column, but leave each rib in place. This procedure will allow you to open wide the thoracic cavity by turning back the flaps of the thoracic wall. Also, turn back the flaps of the abdominal wall. If necessary, wash out coagulated body fluids and surplus preserving fluid.

C. DIGESTIVE AND RESPIRATORY ORGANS OF THE THORAX

The thoracic and abdominal inner organs (or viscera) are situated within the body cavities, or **coelomic cavities,** which collectively are also called the **coelom.** Notice that the body cavities are completely lined with a shiny coelomic epithelium, the **serosa,** which also envelops all the viscera. In life, the serosa secretes a lubricating fluid, which facilitates the smooth gliding of the inner organs relative to one another and to the body wall. This lubrication is necessary because most of the inner organs change their shape and volume rhythmically (e.g., heart, lungs) or cyclically (e.g., stomach, intestine, uterus).

The thoracic part of the coelom comprises two lateral **pleural cavities,** which contain the **lungs** and, between them, the **pericardial cavity** and the **heart** (see Fig. 3-4). The serosa lining the pleural cavities is called **pleura.** The **parietal pleura** lines the wall of the thorax and forms the membranous medial wall of each pleural cavity. The parietal pleurae of the medial wall of each pleural cavity enclose between them a space, the **medi-**

Digestive and Respiratory Organs of the Thorax

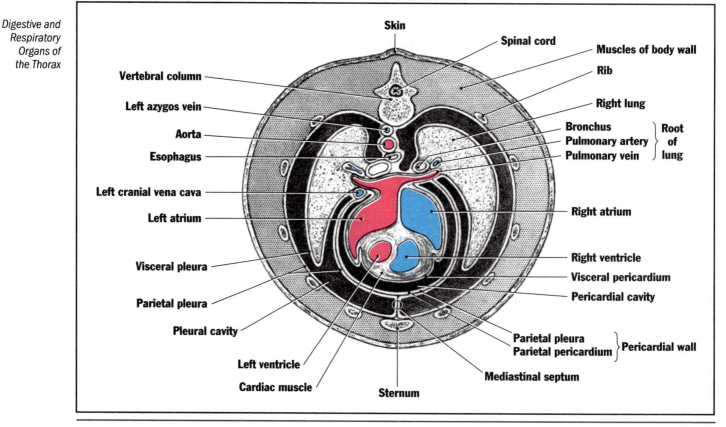

FIGURE 3-4
Diagrammatic cross section through the thorax of a rat at a level near the cranial end of the heart, showing the relationships of the pleural and pericardial cavities and their serosae to the thoracic viscera. The serosa is depicted relatively much thicker than it is in reality in order to show its continuity and anatomical relationships.

astinum, which is filled with connective tissue, major blood vessels, the esophagus, the trachea, and the pericardial cavity with the heart. Ventrally and caudally to the pericardial cavity, the medial parietal pleurae of each pleural cavity touch each other directly and fuse to form the **mediastinal septum.** The **visceral pleura** envelops the lungs and the **roots of the lungs,** which comprise the bronchi and the pulmonary blood vessels supplying the lungs.

The serosa of the pericardial cavity is called the **pericardium.** The **parietal pericardium** lines the wall of the pericardial cavity, and the **visceral pericardium** envelops the heart. The parietal pericardium is fused to the medial parietal pleura of the pleural cavities. The fused parietal pericardium and pleura form the **pericardial wall** of the pericardial cavity.

If your specimen is a young animal, you may see the **thymus** lying across the cranial end of the heart within the mediastinum and extending cranially into the base of the neck (see Fig. 3-5). The thymus has a crucial role in the development of the immune system in mammals by being involved in the maturation of specialized white blood cells (T-lymphocytes), which are responsible for fighting viral, bacterial, and fungal infections. The thymus atrophies after an individual has reached sexual maturity, at which time the maturation process of the T-lymphocytes is concluded. Remove the thymus, being careful not to injure the underlying large blood vessels.

Trace the trachea and esophagus into the thorax, being careful not to injure the artery and vein that lie beside them. Dorsal to the blood vessels entering and leaving the heart, the trachea disappears to divide into **bronchi,** which enter the lungs accompanied by the pulmonary blood vessels (Fig. 3-4). You will be able to see the bronchi later, after the heart has been removed. Notice that the lungs are asymmetrical. The left lung consists of a single large lobe, but the right lung is divided into four lobes: cranial, middle, caudal, and accessory. The accessory lobe passes dorsal to a large vein entering the heart, the **caudal vena cava,** to enter a special pocket of the right pleural cavity, which is formed by the **caval fold,** a mesentery of the caudal vena cava (Fig. 3-5). As a result, the caudal vena cava passes between the caudal and accessory lobes of the right lung.

Cut off a piece of one of the lung lobes and notice its spongy consistency. The bronchi subdivide repeatedly into smaller and smaller passages until they become clusters of thin-walled, microscopic **alveoli,** where the gas

Digestive and Respiratory Organs of the Thorax

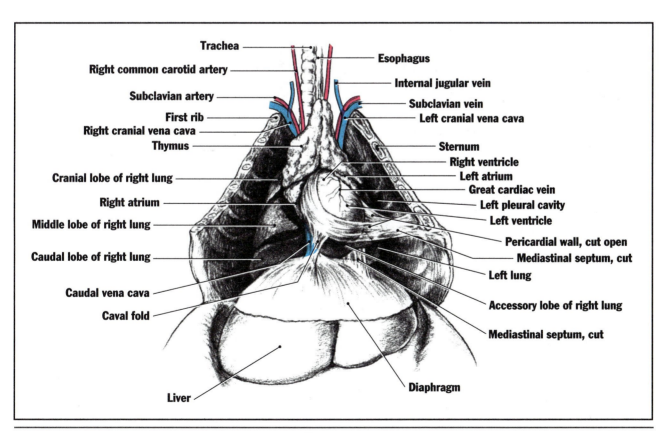

FIGURE 3-5
Ventral view of the organs of the thoracic cavity of a rat.

exchange between the inspired air and the blood in the capillaries in the alveolar wall takes place. Air is moved in and out of the lungs by changes in the size of the pleural cavities. These are enlarged when the diaphragm contracts and pushes the abdominal viscera caudally and when the ribs are rotated cranially and away from the vertebral column during inspiration. Because the lungs keep in contact with the wall of the thorax, they must expand with the pleural cavities, and air rushes into the lungs through the trachea and bronchi as the alveoli expand. The pleural cavities return to their original size when the diaphragm relaxes, the abdominal viscera return to their cranial position, and the ribs are rotated caudally and closer to the vertebral column during expiration. The increased pressure on the lungs and the elastic recoil of the lungs themselves force the air out of the alveoli.

Pull the lungs and heart ventrally and to the specimen's right side, and look into the dorsal part of the left pleural cavity. Observe the esophagus within the mediastinum, dorsal to the heart and lungs. The thin strand accompanying it is the **vagus nerve** (Exercise 6), a cranial nerve that supplies the viscera. Try to follow the esophagus to the stomach by tracing it dorsal to the lungs, heart, and the left part of the liver after you have made a radial cut into the diaphragm up to the esophagus.

*Digestive
Organs of
the Abdomen*

D. DIGESTIVE ORGANS OF THE ABDOMEN

The part of the coelom caudal to the diaphragm is the **peritoneal cavity,** and the serosa in this region is referred to as **peritoneum.** The **parietal peritoneum** lines the body wall, and the **visceral peritoneum** envelopes the abdominal organs. Notice that the organs are connected to each other and with the body wall, especially dorsally, by thin membranes called **mesenteries.** Each mesentery consists of two layers of serosa. The mesenteries suspend and anchor the visceral organs, and it is through them that blood vessels and nerves pass from the body wall to the organs.

The most conspicuous organ in the peritoneal cavity is the **liver,** which fits under the dome of the diaphragm. Lift and push it cranially to get a better view of the **stomach** lying caudal to the liver on the left side of the body (Fig. 3-6). The stomach is a large, bean-shaped sac in which food accumulates before continuing in small portions on its way through the intestine. In the stomach, salivary enzymes continue to act on the food for a while, but the gastric secretion containing hydrochloric acid and the enzyme pepsin soon stops this activity and begins to break down proteins and, in particular, collagen of connective tissue in meat.

The part of the stomach adjacent to the esophageal entrances is its **cardiac region,** the blind saccular part extending to the left is its **fundus,** most of the right side of the stomach is its **body,** and the thick-walled part just cranial to its junction with the small intestine is its **pyloric region** (Fig. 3-7). The **pylorus** is a muscular sphincter between the stomach and the small intestine. The short cranial margin of the stomach between the entrance of the esophagus and the pylorus is its **lesser curvature;** the longer caudal margin is its **greater curvature.** A mesentery known as the **lesser omentum** extends from the lesser curvature of the stomach and small intestine to the liver. The liver, in turn, is connected to the ventral body wall and diaphragm by another mesentery, the **falciform ligament.** Still another mesentery, the **greater omentum,** forms an empty sac that extends caudally from the greater curvature of the stomach and then curves dorsally and folds back cranially to attach to the duodenum (see later) and the body wall. Between the two layers of the serosa that form the mesentery, fat is deposited along the blood and lymphatic vessels creating a network within the greater omentum. The dark, elongated **spleen** is enclosed between the two layers of serosa of the ventral portion of the greater omentum to the left of the stomach. It produces lymphocytes and is a part of the immune system. It also captures and breaks down senescent red and white blood cells. The end products of this breakdown reach the liver, where they are eliminated with the bile.

Cut open the stomach and wash out its contents. The wall of the fundus is thin, and its lining is smooth or thrown into delicate folds. Most of the body and pyloric region of the stomach have a thicker wall, and their lining forms more conspicuous longitudinal folds. The glandular epithelium, which secretes hydrochloric acid and pepsin, is limited to the body and pyloric region of the stomach. The pyloric region ends in the **pylorus,** a muscular sphincter that is closed most of the time, consequently, food is kept in the stomach until muscular churning and chemical action have broken it down enough to be processed by the intestine. A less conspicuous cardiac valve between the stomach and the esophagus normally prevents food from going back into the esophagus.

The first part of the small intestine, the **duodenum,** is C-shaped. As it leaves the stomach, it first extends toward the right side of the body, then turns caudally to about the middle of the peritoneal cavity, and finally turns back medially and cranially nearly up to the level of the stomach (Fig. 3-6). The duodenum is considered to end at the next caudal turn of the small intestine. The rest of the small intestine constitutes the **jejuno-ileum.** Follow it until it joins the large intestine. The small intestine is a long, convoluted organ suspended by a mesentery, which is simply called the **mesentery.** Cut open a part of the small intestine, wash it out, and notice

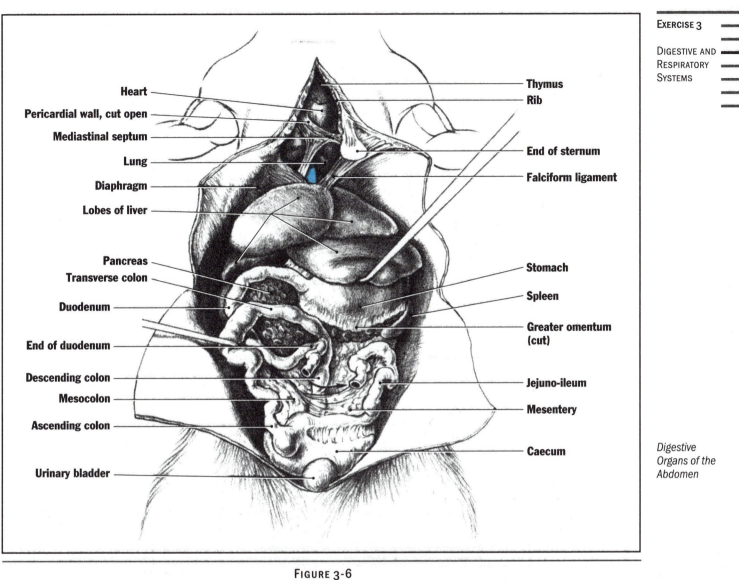

| Heart |
| Pericardial wall, cut open |
| Mediastinal septum |
| Lung |
| Diaphragm |
| Lobes of liver |
| Pancreas |
| Transverse colon |
| Duodenum |
| End of duodenum |
| Descending colon |
| Mesocolon |
| Ascending colon |
| Urinary bladder |

Thymus
Rib
End of sternum
Falciform ligament
Stomach
Spleen
Greater omentum (cut)
Jejuno-ileum
Mesentery
Caecum

Digestive Organs of the Abdomen

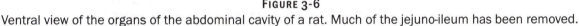

FIGURE 3-6
Ventral view of the organs of the abdominal cavity of a rat. Much of the jejuno-ileum has been removed.

the velvety texture of its lining, which is produced by minute projections, the **villi.** The openings of microscopic intestinal glands lie between the bases of the villi, and the enzymes secreted by these glands, together with those of the pancreas, complete the breakdown of food except for the plant fiber called cellulose. The absorptive surface area of the small intestine is greatly increased by the villi.

The **pancreas** is a diffuse glandular organ. Most of it lies in the dorsal portion of the greater omentum between the duodenum and the spleen. To see the extensive pancreas, turn the stomach in such a manner that its greater curvature points cranially. Most of the dorsal side of the stomach is covered by the pancreas. Try to grab the pancreas with a pair of blunt forceps and pull it away from the stomach. This procedure will show that the greater omentum is a continuous sheet between the

greater curvature of the stomach and the dorsal body wall, and that the spleen, pancreas, and duodenum are built into it.

The enzymes of the pancreas leave through a network of pancreatic ducts, some of which enter the duodenum directly, whereas others join the bile duct (see later).

Finding these minute pancreatic ducts by gross dissection is impractical. The pancreas also secretes bicarbonate ions, which neutralize the strongly acidic food coming from the stomach. Pancreatic enzymes need a neutral environment in which to act. They act on all major categories of food: carbohydrates, proteins, lipids, and nucleic acids. The pancreas also contains microscopic patches of endocrine tissue, the **islets of Langerhans,** which produce insulin and glucagon, hormones that are released into the circulatory system and regulate

45

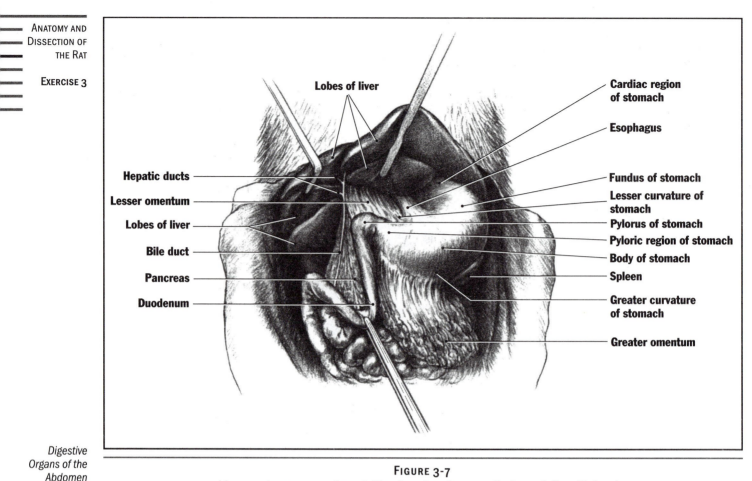

FIGURE 3-7
Liver and pancreas of a rat. The liver has been pulled cranially with hooks
and the duodenum pulled caudally and turned to show the bile duct.

*Digestive
Organs of the
Abdomen*

the blood glucose levels as well as the cellular metabolism of glucose.

The **liver** lies caudal to the diaphragm and is slightly displaced to the right side of the body. It is subdivided into a number of lobes. Push the liver toward the diaphragm, and pull the stomach and duodenum slightly toward the specimen's left. A long whitish duct in the lesser omentum extends from the liver across the dorsal surface of the pancreas toward the duodenum (Fig. 3-7). This is the **bile duct** (*ductus choledochus*). Trace it toward the liver and notice that it is formed by the confluence of several **hepatic ducts** that drain the lobes of the liver. Carefully pick away the pancreatic tissue and trace the bile duct to its entrance into the duodenum at a point about an inch distal to the pylorus. You may notice several minute pancreatic ducts joining the bile duct as you follow it through the pancreas.

Bile secreted by liver cells is discharged into the duodenum through the bile duct. It contains no digestive enzymes, but it is quite alkaline and neutralizes the acidic contents of the stomach. Bile contains bile salts that emulsify fat within the generally watery mixture of food and secretions from the digestive system to facilitate the enzymatic breakdown of fat by the water-soluble lipase from the pancreas. In this manner, the bile salts promote the absorption of fats and fat-soluble vitamins by the small intestine. Bile also contains bile pigments as a waste product from the breakdown of hemoglobin in the liver; they give the bile and the feces their characteristic color.

Unlike mice and most other vertebrates, rats do not have a gallbladder. Bile therefore cannot be stored and concentrated to the extent that it is in most other vertebrates. In addition to producing bile, the liver has a vital function in the storage and metabolism of food products brought to it by the veins draining the stomach and intestines (Exercise 4).

Find the point at which the jejuno-ileum enters a large, blind sac called the **caecum** (Figs. 3-6 and 4-3, Exercise 4). This part of the large intestine is quite long in rodents and many other herbivores, such as horses and rabbits, for its function is similar to that of the rumen in cattle: It contains symbiotic bacteria and protozoans, some of which produce the enzyme cellulase, which breaks down the plant product cellulose by fermentation. Many of the microorganisms, which reproduce rapidly, are themselves digested and represent a source of proteins for the host. Most of the fatty acids, amino acids, acetic acid, and other breakdown products from the fermentation process are absorbed by the caecum. A small vermiform appendix at the end of the caecum is present in many mammals, but it is absent in rats.

Unabsorbed material from the caecum moves into the **colon,** which leaves the caecum at the site immediately adjacent to the entrance of the jejuno-ileum (Fig. 3-6). Cut open the caecum on the side opposite this junction, wash out its contents, and notice the common opening of the jejuno-ileum and colon. A sphincter valve at the end of the small intes-tine prevents the caecal contents from backing up. Also notice that villi are absent in the caecum. The colon, the diameter of which is not much larger than that of the small intestine, ascends to the level of the stomach, crosses toward the left side of the body, and descends against the back. The mesentery that anchors the colon to the dorsal body wall is called **mesocolon** (Fig. 3-6). The colon often has a beaded appearance because of the fecal pellets it contains. It extends into the pelvic canal dorsal to the reproductive organs and enters a short **rectum,** which terminates at the **anus.** You will observe the rectum later (Exercise 5). Remaining products of caecal digestion, water, and some vitamins are absorbed by the colon. Undigested residue and excess bacteria are expelled as the feces.

*Digestive
Organs of the
Abdomen*

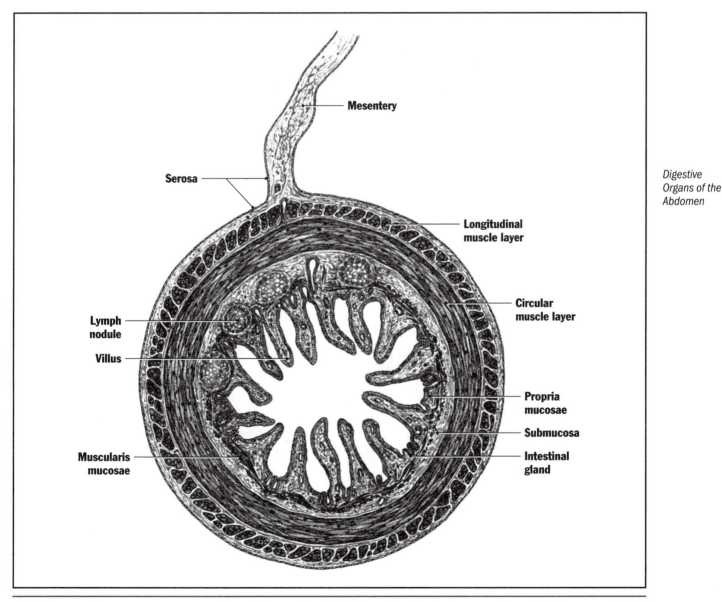

Mesentery

Serosa

**Longitudinal
muscle layer**

**Circular
muscle layer**

**Lymph
nodule**

Villus

**Propria
mucosae**

Submucosa

**Intestinal
gland**

**Muscularis
mucosae**

FIGURE 3-8
Microscopic cross section through the ileum of a mammal.

E. MICROSCOPIC STRUCTURE OF THE SMALL INTESTINE

Study a histological slide preparation of a cross section through the small intestine of a mammal, and compare it with Figs. 3-8, 3-9 and 3-10. Beginning at the outside and progressing toward the lumen, the following layers should be noted: (1) serosa, (2) longitudinal muscle layer, (3) circular muscle layer, (4) submucosa, and (5) mucosa.

The **serosa** is a thin layer of simple squamous epithelium (the visceral peritoneum) supported by connective tissue and containing a few blood vessels. About all that can be seen of the epithelial cells are their flattened nuclei. The attachment of the mesentery to the intestine may be seen in some slides.

The **longitudinal muscle layer** is composed of **smooth muscle** fibers that spiral around the intestine in a plane close to its longitudinal axis. Because these fibers are overlapping, spindle-shaped cells being viewed primarily in cross section, their diameters vary considerably, as can be seen under high magnification. Smooth muscle fibers in the **circular muscle layer** form a tight spiral whose coils lie close to the transverse plane of the intestine, so they are seen in oblique to longitudinal sections. Under high magnification, each of these fibers can be seen to contain a single, elongate nucleus, located near its center, and many longitudinal nonstriated **myofibrils.**

These two muscle layers, longitudinal and circular, function as antagonists to each other and are responsible for the churning movements that mix food and enzymes and for the peristaltic movements that move the food down the intestine. They are innervated by the autonomic nervous system; the parasympathetic portion of the autonomic system provides activating stimuli and the sympathetic portion provides inhibiting stimuli. The cell bodies of the postganglionic parasympathetic neurons are concentrated in more lightly stained **intramural ganglia** within the wall of the digestive tract, either between the longitudinal and circular muscle layers or between the circular muscle layer and the submucosa.

The **submucosa** consists of loose, fibrous, connective tissue containing many blood vessels. If your histological slide has been taken near the colic end of the small intestine, large, deep-staining masses of lymphocytes called **lymph nodules** can be seen along one side of the intestine, lying in the submucosa and extending into the mucosa. As mentioned earlier, T-lymphocytes mature in the thymus. It is probable that a second type, B-lymphocytes, mature in lymphoid tissue in the wall of the digestive tract before being distributed to other lymphoid tissues. On exposure to foreign antigens, primarily of bacterial origin, certain B-lymphocytes become antibody-producing cells.

The **mucosa** is the most complex layer of the intestine and can be subdivided into three layers (Figs. 3-8, 3-9, and 3-10). The **muscularis mucosae** is a thin layer of longitudinal and circular smooth muscle fibers and lies internally next to the submucosa. The **propria mucosae,** the layer of connective tissue next to the muscularis mucosae, contains blood and lymphatic vessels. The **mucosal epithelium** lines the intestinal lumen; it is a simple columnar epithelium and comprises many lightly stained **goblet cells.** These cells secrete mucus, which

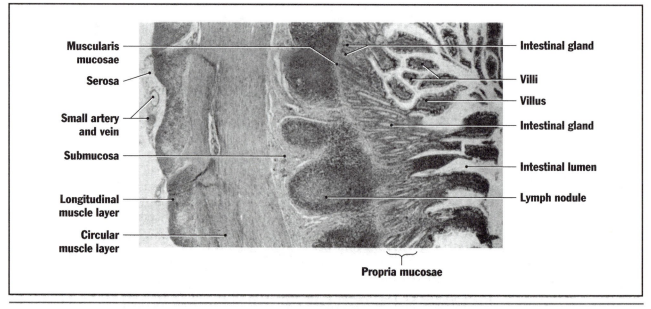

Muscularis mucosae
Serosa
Small artery and vein
Submucosa
Longitudinal muscle layer
Circular muscle layer
Intestinal gland
Villi
Villus
Intestinal gland
Intestinal lumen
Lymph nodule
Propria mucosae

FIGURE 3-9
Photomicrograph of a transverse section through the wall of the ileum of a cat.

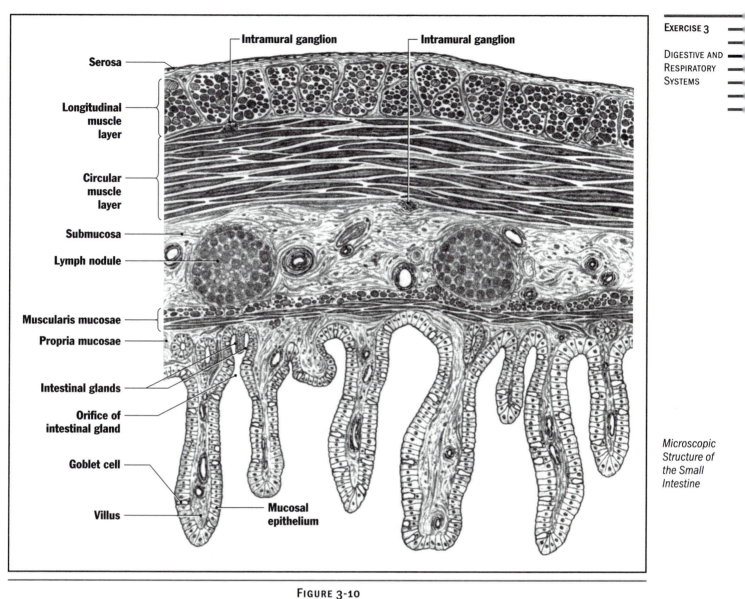

Intramural ganglion

Intramural ganglion

Serosa

Longitudinal
muscle
layer

Circular
muscle
layer

Submucosa

Lymph nodule

Muscularis mucosae

Propria mucosae

Intestinal glands

Orifice of
intestinal gland

Goblet cell

Villus

Mucosal
epithelium

*Microscopic
Structure of
the Small
Intestine*

FIGURE 3-10
Diagram of an enlargement of part of a cross section through the wall of the ileum of a mammal.

lubricates the food within the intestinal lumen and helps protect the delicate intestinal lining, especially from the effects of proteolytic enzymes.

The numerous, fingerlike projections that extend into the lumen of the intestine are called **villi** (Fig. 3-10). The core of each villus is formed by the propria mucosae and some muscle fibers that extend from the muscularis mucosae. The latter convey motility to the villi. The villi and the ultramicroscopic **microvilli** on the surface of the epithelial cells greatly increase the absorptive surface area of the small intestine.

Between the villi, the mucosal epithelium dips deeply into the propria mucosae and forms the tubular **intestinal glands.** Because of the three-dimensional arrangement of these glands, your histological slide preparation may show the intestinal glands and villi only in cross or oblique sections. The intestinal glands and the columnar cells of the mucosal epithelium secrete intestinal enzymes, some of which are released into and act within the lumen of the intestine. Other intestinal enzymes remain associated with the microvilli at the surface of the epithelial cells.

EXERCISE

FOUR

Circulatory System

T HE CIRCULATORY SYSTEM is the transport system of the body. It carries substances absorbed by the digestive system and oxygen from the lungs to all body cells, and it carries waste products of cellular metabolism to organs that eliminate them. Carbon dioxide is eliminated by the lungs and nitrogenous wastes by the kidneys. Substances enter and leave the body by diffusion, supplemented in some cases by active transport. The circulatory system, in essence, shortens diffusion distance by providing a mechanism for the bulk flow of materials between sites of entry into the body, sites of consumption (the cells), and sites of waste removal. Bulk flow allows animals to grow larger than a few millimeters in diameter and to lead active lives. The circulatory system also helps to distribute heat throughout the body, transports hormones, helps to maintain a constant internal environment, and defends the body from disease organisms.

A. PATTERN OF CIRCULATION

The blood vessels to be exposed will mean more to you if, before dissecting them, you understand the basic pattern of circulation. Blood is transported in **arteries** from the heart to minute, thin-walled **capillaries** in the tissues, where exchanges of water, nutrients, gases, and excretory products between the blood and interstitial fluid, which bathes the cells, takes place. The capillary beds are drained primarily by **veins,** which return to the heart, but some surplus interstitial fluid and a few protein molecules that seep out of the vascular capillaries return to the larger veins in **lymphatic vessels.** Lymphatic vessels are difficult to see in gross dissections, although some of the lymph nodes that lie along their course will be noticed. **Lymph nodes** are sites for the production of certain white blood cells, some of which can respond to foreign antigens by transforming themselves into antibody-producing cells. Other cells in lymph nodes help to destroy bacteria and other foreign cells that enter the body by ingesting and digesting them.

Unlike the open circulation of many invertebrates, in which all components of the blood pass freely among the cells of the body, the circulatory system of vertebrates is a closed system. Most components of the blood remain in the vessels and, for the most part, only small molecules are exchanged in the capillary beds with the interstitial fluid. Although an open system is well adapted to the mode of life of many invertebrates, a closed system allows for higher blood pressures and a more rapid distribution of materials; the closed system also allows, by dilation and constriction of selective vessels, for control over the amount of blood going to different regions of the body. Active organs receive more blood than others.

In an adult mammal (Fig. 4-1), blood low in oxygen content and high in carbon dioxide is returned from the body to the **right atrium** of the heart by two or three major veins and their tributaries. In human beings and many other mammals, a single **cranial** (superior) **vena cava** drains the head, neck, and arms, but rats have separate left and right cranial venae cavae. In all mammals, a single **caudal** (inferior) **vena cava** drains the caudal parts of the body. Blood returning from the stomach and intestines is carried first to capillary-like spaces in the liver (the **hepatic sinusoids**) by the **hepatic portal vein.** The liver drains into the caudal vena cava through many **hepatic veins.**

As blood low in oxygen content and high in carbon dioxide enters the right atrium, blood that has gained oxygen and lost carbon dioxide in the capillaries of the lungs returns to the **left atrium** of the heart in **pulmonary veins.** (Notice that veins are blood vessels that lead to the heart, regardless of the oxygen content of their blood. Similarly, arteries always lead away from the heart.) When the ventricles expand after a contraction,

and the right and left atria contract, which they do simultaneously, blood within them is discharged into the respective ventricles. When the thick, muscular walls of the ventricles contract, the increasing pressure closes the atrioventricular valves, and blood is driven with considerable force into the two large arteries leaving the heart. Closure of valves at the base of each of these vessels prevents a backflow of blood into the ventricles as they relax to be refilled from the atria. A **pulmonary trunk** leaves the **right ventricle** and soon branches into a pair of **pulmonary arteries** that carry oxygen-depleted blood to the lungs, where gas exchange takes place. An **aorta** leaves the **left ventricle,** arches to the left side of the body, and descends to the pelvis. On its way, it gives off numerous branches that deliver oxygen-rich blood to all parts of the body.

Blood flowing through the heart does not supply the heart musculature itself with the oxygen and nutrients that it needs. A separate system of **coronary arteries** leaves the base of the aorta and supplies capillaries in the heart wall. **Cardiac veins** drain the heart wall and return oxygen-depleted blood to the base of the left cranial vena cava.

Both the arteries and the major veins should be injected in your specimen, but valves in the veins sometimes prevent a thorough penetration of the substances with which they have been injected. If the veins are not injected, they will appear as thin-walled, translucent tubes in which a bit of blood or preserving fluid can be seen. Because most veins lie beside the corresponding arteries, it will be convenient to study the arteries and veins together in many regions of the body. Do not be surprised to find variation among individual specimens in the locations at which smaller vessels join the main ones. Often the only way to identify a vessel is to trace it to the organ it supplies or drains.

B. THE HEART AND ITS GREATER VESSELS

Before examining the heart and the associated vessels, you should find and trace superficial veins in the neck toward the heart. Some may have been partly destroyed when you studied the muscles and organs in the neck (Exercises 2 and 3), but examine their pattern in Fig. 4-2 (right side of figure) and find as many as you can.

A pair of relatively large **external jugular veins** lie superficially on the lateroventral surface of the neck. One was probably used to inject veins. They drain most of the head. Each external jugular vein extends caudally, crosses the ventral surface of the clavicle, and immediately is joined by an **axillary vein** to form the **subclavian vein.** The axillary vein drains the armpit (axilla), shoulder, and

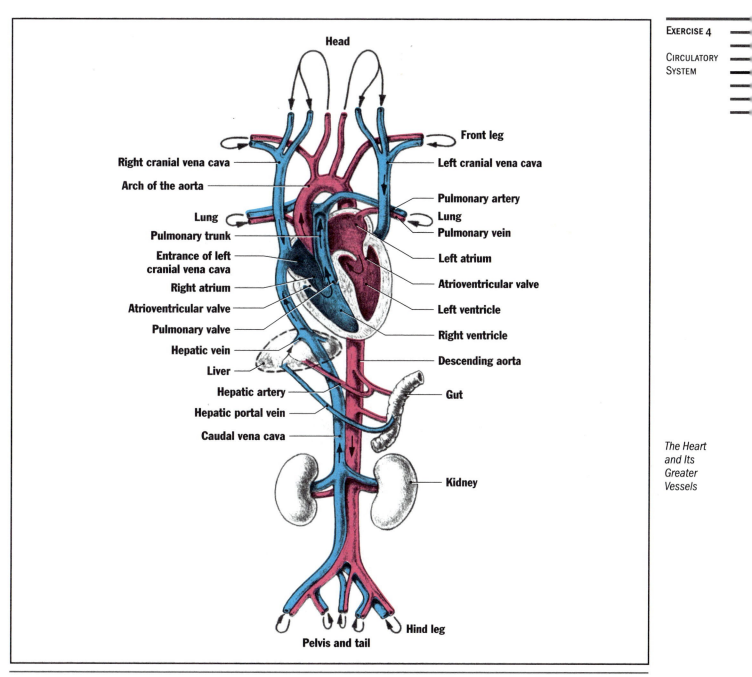

Head

Front leg

Right cranial vena cava ——— ——— Left cranial vena cava

Arch of the aorta ———

——— Pulmonary artery

Lung ——

——— Lung

Pulmonary trunk ———

——— Pulmonary vein

Entrance of left
cranial vena cava ———

——— Left atrium

Right atrium ———

——— Atrioventricular valve

Atrioventricular valve ———

——— Left ventricle

Pulmonary valve ———

——— Right ventricle

Hepatic vein ———

——— Descending aorta

Liver ———

Hepatic artery ———

——— Gut

Hepatic portal vein ———

Caudal vena cava ———

——— Kidney

Hind leg

Pelvis and tail

*The Heart
and Its
Greater
Vessels*

FIGURE 4-1
Diagram of the basic pattern of blood circulation in an adult rat. This is a
ventral view, so the animal's right side is on the left side of the drawing.

arm. It is often not well injected, but look for it where it crosses superficially to the first rib and joins the external jugular vein posteriorly and laterally. One of its most conspicuous branches lies on the lateral surface of the thorax. The subclavian vein then turns deeply, passing between the clavicle and first rib.

To trace the subclavian vein, cut across the muscles attaching to the front of the sternum (sternomastoid and cleidomastoid), the clavicle near its attachment to the sternum on both sides (if this has not already been done), the first rib on each side, and transversely across the sternum between the first two ribs. Remove the piece of sternum you have freed. As each subclavian vein curves caudally deep to the first rib, it receives a small **internal jugular vein** that helps to drain the inside of the skull and descends beside the trachea. The confluence of the subclavian and internal jugular veins forms the **cranial** (superior) **vena cava.** Rats have both a left and right cranial vena cava. Each cranial vena cava extends caudally into the thoracic cavity.

53

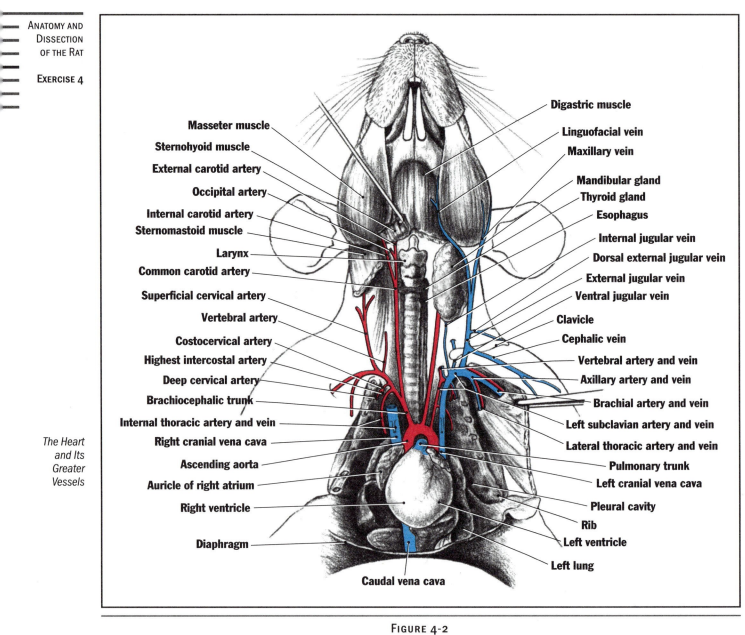

Masseter muscle
Sternohyoid muscle
External carotid artery
Occipital artery
Internal carotid artery
Sternomastoid muscle
Larynx
Common carotid artery
Superficial cervical artery
Vertebral artery
Costocervical artery
Highest intercostal artery
Deep cervical artery
Brachiocephalic trunk
Internal thoracic artery and vein
Right cranial vena cava
Ascending aorta
Auricle of right atrium
Right ventricle
Diaphragm

Digastric muscle
Linguofacial vein
Maxillary vein
Mandibular gland
Thyroid gland
Esophagus
Internal jugular vein
Dorsal external jugular vein
External jugular vein
Ventral jugular vein
Clavicle
Cephalic vein
Vertebral artery and vein
Axillary artery and vein
Brachial artery and vein
Left subclavian artery and vein
Lateral thoracic artery and vein
Pulmonary trunk
Left cranial vena cava
Pleural cavity
Rib
Left ventricle
Left lung
Caudal vena cava

The Heart and Its Greater Vessels

FIGURE 4-2
Ventral view of the cranial arteries and veins of the rat. The clavicle, superficial
neck muscles, and veins have been removed on the left side of the drawing.

Spread the walls of the thoracic cavity apart. The heart and the associated blood vessels may still be covered by the thymus. If so, carefully remove it. If the pericardial sac surrounding the heart has not been cut open yet, do so now. Then cut the pericardial sac away from its attachment to the vessels entering and leaving the heart. Pick away connective tissue cranial to the heart and trace each cranial vena cava toward the heart. A group of arteries lies between them (Fig. 4-2).

The heart consists of **right** and **left ventricles,** which are not easy to distinguish externally, and **right** and **left atria,** whose ear-shaped, dark-colored **auricles** on each side of the heart are readily apparent in a ventral

view. The **caudal** (inferior) **vena cava,** which you have already identified (Exercise 3), returns blood from the caudal part of the body to the right atrium. The right and left cranial vena cava also enter the right atrium. You must lift up the caudal end of the ventricles to see the left cranial vena cava crossing the dorsal surface of the heart to enter. Several **cardiac veins,** which drain the heart wall, will be seen joining the base of the left cranial vena cava. This part of the left cranial vena cava could be called the **coronary sinus,** because it is comparable to the coronary sinus of human beings.

Oxygen-depleted blood goes from the right atrium into the right ventricle, which pumps it to the lungs through the **pulmonary trunk.** The pulmonary trunk leaves the cranial end of the right ventricle and curves dorsally. Carefully dissect between it, the left auricle, and the left cranial vena cava to find the place at which it bifurcates into two **pulmonary arteries,** which continue to the lungs. The **pulmonary veins** returning oxygen-rich blood to the left atrium lie so deep that you will not be able to see them until you remove the heart at the end of this exercise.

The first part of the aorta, carrying oxygen-rich blood from the **left ventricle** to the body, is known as the **ascending aorta** because it curves cranially. It first goes deep to the pulmonary trunk. Push the right auricle aside and find the base of the aorta. Its first branch is a small **coronary artery** to the heart wall. Another coronary artery branches from the other side of the aorta's base, but it is difficult to find.

C. VEINS AND ARTERIES OF THE THORAX, NECK, AND HEAD

Follow the ascending aorta and the paired cranial venae cavae lying on each side of it (Fig. 4-2). After a short distance, the aorta arches to the left, goes deep to the left cranial vena cava, and descends through the thorax. Three branches leave the **arch of the aorta.** From the specimen's right to left, they are the **brachiocephalic trunk,** the **left common carotid artery,** which extends cranially deep to the sternohyoid muscle and lateral to the trachea to supply the left side of the head, and the **left subclavian artery,** which extends cranially and laterally toward the left shoulder and arm. The brachiocephalic trunk extends cranially about one centimeter and then divides into the **right common carotid** and **right subclavian arteries.** We will return to these branches of the aorta later.

Look dorsal to the left lung. Most of the intercostal spaces between the ribs are supplied by paired **intercostal arteries** that leave the **descending aorta,** and most are drained by **intercostal veins** that enter a **left azygos (hemiazygos) vein** lying beside the aorta. Blood in the azygos vein flows cranially to enter the left cranial vena cava. There is no right azygos vein in rats and mice, because the right intercostal veins cross the aorta dorsally to enter the left azygos vein.

Now trace one of the subclavian arteries and the accompanying cranial vena cava and its tributaries. Select the side in which the cranial vena cava is better injected. The arteries will be easier to find than the veins, but as you identify an artery, look for the accompanying vein. The first branch of a subclavian artery is an **internal thoracic artery,** which leaves the ventral surface of the subclavian artery and passes to the inside of the thoracic wall beside the sternum. An **internal thoracic vein**

accompanies the artery and enters the cranial vena cava. Cut these vessels near the thoracic wall.

Three other arteries leave the subclavian artery nearby. A **costocervical trunk** leaves the dorsal surface of the subclavian artery nearly opposite the origin of the internal thoracic artery and extends caudally and laterally. Its major branches are the **highest intercostal artery,** which supplies the first two or three intercostal spaces, and a **deep cervical artery,** which supplies the muscles situated deeply at the base of the neck. The **highest intercostal** and **deep cervical veins** usually converge to form a **costocervical vein** before entering the cranial vena cava, but it is not uncommon for them to join the cranial vena cava independently.

The **vertebral artery** leaves the subclavian artery slightly cranial to the internal thoracic artery and costocervical trunk. It extends cranially and deeply to enter the transverse canals of the cervical vertebrae, through which it runs to the ventral surface of the brain. A **vertebral vein** accompanies it and enters the cranial vena cava.

The **superficial cervical artery,** which is not accompanied by a comparable vein, next leaves the subclavian artery and passes cranially and laterally deep to the clavicle to supply superficial neck muscles and certain shoulder muscles. After giving off the superficial cervical artery, the subclavian artery curves in front of the first rib to enter the armpit. At this point, its name is changed to **axillary artery,** for it continues laterally through the axilla. You may have found the axillary vein, which lies beside the axillary artery. Recall that the union of the axillary and external jugular veins form the subclavian vein, which leads to the cranial vena cava.

A small **ventral** (anterior) **jugular vein,** which is easily overlooked, enters the subclavian vein, or sometimes the external jugular vein, from the ventral surface of the neck.

The axillary artery gives rise to the **lateral thoracic artery,** which supplies the lateral surface of the thorax, and then continues into the arm as the **brachial artery. Lateral thoracic** and **brachial veins** accompany these arteries.

Again find the external jugular vein and trace it cranially. As it crosses the clavicle, it receives a **cephalic vein** draining the lateral surface of the arm and adjacent shoulder muscles. Farther rostrally, the external jugular vein receives the **dorsal** (posterior) **external jugular vein,** which drains chiefly the trapezius muscles on the laterodorsal side of the neck and head. The external jugular vein bifurcates near the caudal border of the mandibular gland. Its ventral branch, the **linguofacial vein,** drains the tongue, the lower jaw, and the front of the face; its dorsal branch, the **maxillary vein,** drains the more dorsal and caudal parts of the head and face and receives several vessels from within the skull.

Veins and Arteries of the Thorax, Neck, and Head

Return to the base of the neck and trace one of the common carotid arteries cranially. It parallels the trachea and at the level of the thyroid gland bifurcates into the **external** and **internal carotid arteries.** The external carotid divides almost immediately into branches that supply the tongue, jaw muscles, eye, and other structures on the outside of the head. The internal carotid artery extends deeply toward the base of the skull and enters the cranial cavity to help carry blood to the brain. Do not confuse the internal carotid artery with the first branch (**occipital artery**) of the external carotid artery. The occipital artery also extends deeply, but it crosses the internal carotid artery laterally and continues dorsally to supply muscles on the back of the skull and on the dorsal surface of the neck.

The white strand that lies between the common carotid artery and the internal jugular vein is the tenth cranial nerve, the **vagus nerve.** The vagus nerve continues caudally, crosses the ventral surface of the subclavian artery, and then runs dorsally to follow the esophagus to the abdominal viscera (Exercise 3). Look dorsal to the left lung and you will see this part of the esophagus and the accompanying vagus nerve lying between the descending aorta and the heart. Another small nerve, the cervical extension of the **sympathetic trunk,** lies dorsal to the vagus nerve and common carotid artery, but this part of it is difficult to see. It continues caudally into the thorax, where it lies against the chest wall, ventral and lateral to the heads of the ribs. The left sympathetic trunk can be found embedded in the fat dorsal to the azygos vein. The sympathetic trunk is a part of the autonomic nervous system (Exercise 6).

D. VEINS AND ARTERIES OF THE ABDOMEN

D.1. Hepatic Portal System and Hepatic Veins

The abdominal viscera, hind legs, and tail are supplied by branches of the descending aorta. All of them are drained ultimately by the caudal vena cava, but the digestive organs and the spleen are drained directly by a group of veins that first carries blood to the liver. These veins constitute the **hepatic portal system.** A portal system is defined as a group of veins that lead from a capillary bed in one organ to a capillary bed in another organ. Blood from organs drained by the hepatic portal system passes through the hepatic sinusoids within the liver. Because these sinusoids lack the complete lining of endothelial cells, which is present in true capillaries, blood comes into direct contact with hepatic cells. Many metabolic conversions occur in the liver: just after a meal, excess absorbed sugars are stored, largely in the form of glycogen, but between meals, sugar may be released by the liver to keep the glucose content of the

blood constant; amino acids are deaminated and their amino groups are converted into urea; some absorbed toxins are removed or broken down (e.g., alcohol); certain products from the breakdown of red blood cells in the spleen are salvaged, and others are removed and excreted as bile pigments.

Hepatic portal veins are not injected, but the main ones can be seen nevertheless. Tear the ventral wall of the greater omentum, if you have not already done so, pull the stomach cranially and the spleen to the specimen's left, so that you stretch the dorsal wall of the omentum (Fig. 4-3). You will see small arteries going to the spleen, stomach, and duodenum. Carefully pick away pancreatic tissue, and you will find thin-walled veins beside these arteries. A **lienic vein** comes from the spleen, receives a **left gastric vein** from part of the stomach, and soon unites with the large **cranial mesenteric vein,** which drains the **jejunal veins** from the small intestine to form the **hepatic portal vein.** The hepatic portal vein receives a small **gastroduodenal vein** from the pyloric region of the stomach, the proximal part of the duodenum, and adjacent parts of the pancreas. It then passes dorsal to the pylorus and duodenum to enter the lesser omentum, through which it runs cranially to the liver, lying dorsal to the bile duct and a small hepatic artery.

Push the liver caudally and cut through the diaphragm to find the caudal vena cava emerging from the cranial surface of the liver. You may observe several small **phrenic veins** joining the caudal vena cava from the diaphragm. Dissect away liver tissue to locate some of the many **hepatic veins** that drain the liver and enter the caudal vena cava. Notice the site at which the caudal vena cava enters the caudodorsal part of the liver near the large, bean-shaped right kidney. It is important to recognize that, although the caudal vena cava runs through the liver, it is receiving blood from the liver and not carrying blood to this organ.

D.2. Descending Aorta and Posthepatic Caudal Vena Cava

Again find the **descending aorta** in the left pleural cavity. Cut through the diaphragm, noticing the small **phrenic arteries** that supply it, and trace the aorta into the abdominal cavity. It lies to the left of the caudal vena cava. Carefully pick away the connective tissue and fat that covers these large vessels and lies between them and the pair of **kidneys.** The kidneys lie dorsal to the parietal peritoneum, and each is partly covered by a capsule of fat. As you dissect this tissue away, you may notice a tough, whitish strand. This is the **splanchnic nerve,** a branch of the sympathetic trunk that carries sympathetic fibers to the abdominal viscera. It first goes to ganglia located near the base of the coeliac and cranial mesenteric arteries (see later), and from there sympathetic fibers follow the arteries to the visceral organs.

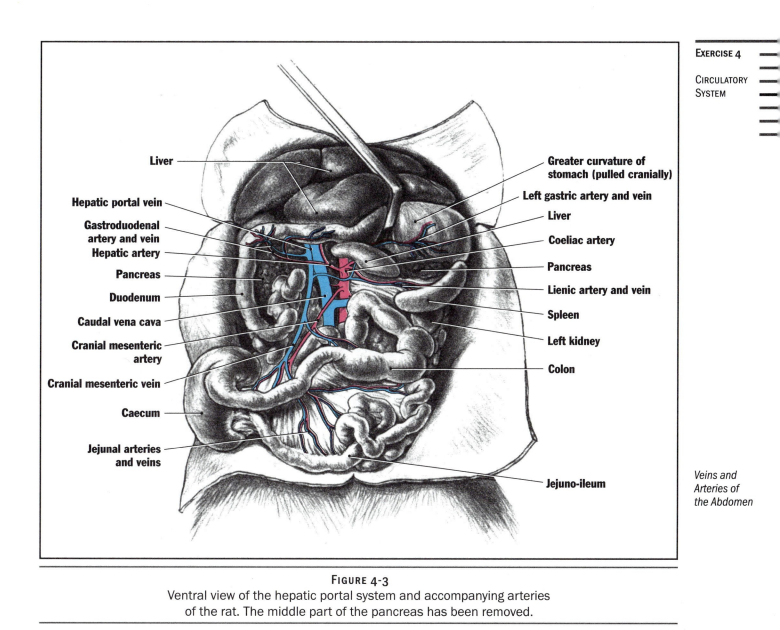

Liver

Hepatic portal vein

Gastroduodenal
artery and vein

Hepatic artery

Pancreas

Duodenum

Caudal vena cava

Cranial mesenteric
artery

Cranial mesenteric vein

Caecum

Jejunal arteries
and veins

Greater curvature of
stomach (pulled cranially)

Left gastric artery and vein

Liver

Coeliac artery

Pancreas

Lienic artery and vein

Spleen

Left kidney

Colon

Jejuno-ileum

*Veins and
Arteries of
the Abdomen*

FIGURE 4-3
Ventral view of the hepatic portal system and accompanying arteries
of the rat. The middle part of the pancreas has been removed.

Notice and save the small, round nodule of glandular tissue, the **adrenal** (suprarenal) **gland,** which lies at the cranial end of each kidney. The adrenal gland is an endocrine gland consisting of two parts. Its medulla secretes the hormone adrenalin (epinephrine), which helps the sympathetic nervous system adjust the body to stress (the flight or fight reaction, Exercise 6). Its cortex secretes three groups of cortical hormones: one helps to regulate salt and mineral metabolism; another regulates carbohydrate and protein metabolism; and the third group influences sexual development. Also save the **ureter,** the duct that drains the kidney and extends caudally into the pelvic canal.

The first branch of the abdominal segment of the aorta is the **coeliac artery,** which leaves the ventral surface of the aorta (Fig. 4-3). It soon divides into three branches: a **lienic artery** to the spleen, a **left gastric artery** to part of the stomach; and a **hepatic artery,** which, after giving off a branch to supply part of the pancreas, duodenum, and stomach, continues to the liver. The hepatic artery delivers oxygen-rich blood to the liver, which is metabolically very active. All these arteries follow some of the hepatic portal veins that you previously observed (Fig. 4-3).

The next branch, the **cranial mesenteric artery,** also leaves the ventral surface of the aorta (Fig. 4-4). Accompanied by the jejunal veins, it goes to most parts of the small intestine and to a part of the large intestine. At approximately the level of the cranial mesenteric artery, paired **renal arteries** branch off the lateral surface of the aorta to the kidneys. **Renal veins** return blood from the kidneys to the caudal vena cava. Small **adrenal arteries** to the adrenal gland leave either the renal arteries or the aorta directly. They are accompanied by **adrenal veins,** which enter the renal veins or the caudal vena cava.

57

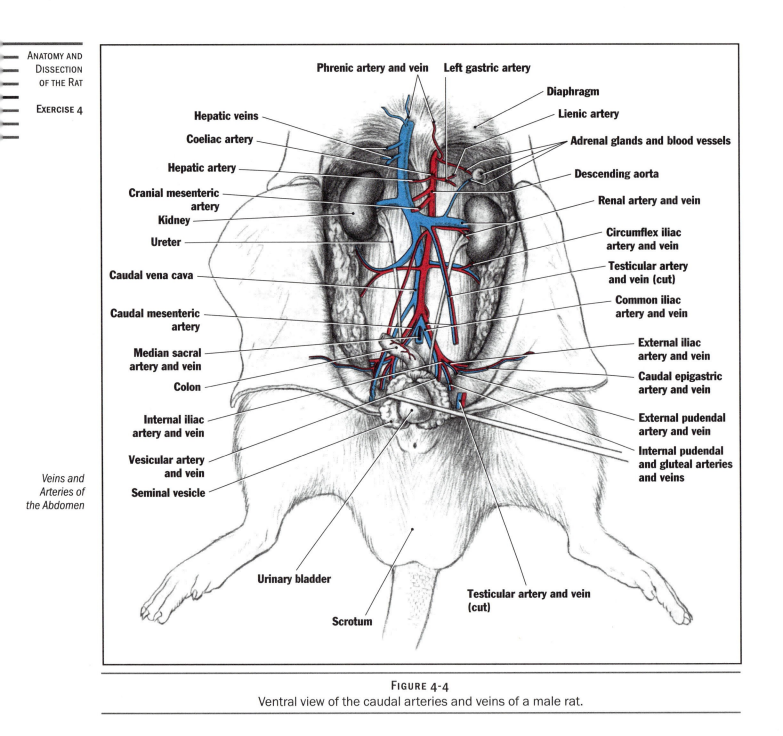

Phrenic artery and vein   **Left gastric artery**

Hepatic veins

Coeliac artery

Hepatic artery

Cranial mesenteric artery

Kidney

Ureter

Caudal vena cava

Caudal mesenteric artery

Median sacral artery and vein

Colon

Internal iliac artery and vein

Vesicular artery and vein

Seminal vesicle

Diaphragm

Lienic artery

Adrenal glands and blood vessels

Descending aorta

Renal artery and vein

Circumflex iliac artery and vein

Testicular artery and vein (cut)

Common iliac artery and vein

External iliac artery and vein

Caudal epigastric artery and vein

External pudendal artery and vein

Internal pudendal and gluteal arteries and veins

Urinary bladder

Testicular artery and vein (cut)

Scrotum

Veins and Arteries of the Abdomen

FIGURE 4-4
Ventral view of the caudal arteries and veins of a male rat.

Depending on the sex of the animal, paired **testicular** or **ovarian arteries** leave the aorta slightly caudal to the renal arteries. The testicular arteries extend directly caudally into the scrotum; you will examine their distribution to the testes later, after you have opened the scrotum. The ovarian arteries extend laterally, just caudal to the kidneys, to supply the ovaries and the uterine horns. **Testicular** or **ovarian veins** accompany the arteries. The right testicular or ovarian vein joins the caudal vena cava, and the left testicular or ovarian vein of most specimens enters the left renal vein.

Paired **circumflex iliac arteries** branch from the aorta farther caudally and help to carry blood to muscles of the back and to the cranial muscles of the pelvis. The back is also supplied by several pairs of **lumbar arteries** that leave the dorsal surface of the aorta. Satellite lumbar veins empty into the caudal vena cava.

Just before the aorta terminates, it gives off a single **caudal mesenteric artery** to the caudal part of the colon and rectum. Veins draining this part of the digestive tract are tributaries of the caudal mesenteric vein, which joins the cranial mesenteric vein and enters the hepatic portal system. At their caudal ends, both the aorta and the caudal vena cava bifurcate to form the **common iliac arteries** and **veins.** These veins lie deep to the arteries and may be difficult to see. Carefully push the reproductive organs aside, lift up one common iliac artery and vein, dissect deep to the termination of the aorta and caudal vena cava, and look for the **median sacral artery** and **vein,** which supply and drain the sacrum and tail.

Follow one common iliac artery and vein as they pass laterally and caudally toward the base of the leg. Near the leg, they bifurcate into **external** and **internal iliac arteries** and **veins.** The external iliac artery and vein pass through the body wall, going deep to a white inguinal ligament. When they enter the hind leg, these vessels are known as the **femoral artery** and **vein.** Two small arteries usually arise from the external iliac artery just before it passes through the body wall. A **caudal epigastric artery** turns onto the inside of the ventral abdominal wall, and an **external pudendal artery** goes to a wad of fat in the groin outside the body wall and to parts of the external genitalia. These arteries usually arise from the external iliac artery by a common trunk, but they may arise independently, and sometimes the external pudendal arises from the femoral artery. Accompanying veins have a similar pattern.

The first branch of the internal iliac artery is the **vesicular artery** to the urinary bladder and adjacent urogenital organs, including the uterus in the female. This artery develops from the large umbilical artery of the embryo, which carries blood to the placenta. It is accompanied by a **vesicular vein,** which enters the internal iliac vein. (The umbilical vein of the embryo, which returns blood from the placenta, enters the liver and does not develop into an adult vessel.) A short distance from the point at which it gives off the vesicular artery, the internal iliac artery divides into a number of arteries that supply pelvic muscles and urogenital organs; among them are the **internal pudendal** and **gluteal arteries.** Veins have a similar pattern.

Refer to Fig. 4-1 to review the overall pattern of circulation.

E. ROOT OF THE LUNG AND INTERNAL STRUCTURE OF THE HEART

Although the heart of the rat is not large, major features can easily be seen. Carefully cut through the descending aorta and the cranial and caudal venae cavae. Leave rather long stumps attached to the heart. Lift up the cranial end, or base, of the heart, dissect beneath it, and gradually pull the heart and attached vessels caudal. The pulmonary arteries, which will be injected in blue

because they received the injection mass from the large veins of the body and the right atrium and ventricle, can now be seen more clearly and traced to each lung. As you trace them, notice that the trachea divides into a pair of **bronchi** going to the lungs (Fig. 3-4). Uninjected **pulmonary veins,** which return oxygen-rich blood from the lungs to the left atrium of the heart, can be seen caudal to the distal part of the pulmonary arteries. The area at which the pulmonary blood vessels and bronchi enter and leave each lung is called the **root of the lung.** Cut through the pulmonary blood vessels near the lungs and remove the heart.

Orient the heart and reidentify the blood vessels that enter and leave it (Figs. 4-2 and 4-5). Venae cavae enter the right atrium. Cardiac veins, which drain the heart wall and enter the base of the left cranial vena cava (coronary sinus), can now be seen more clearly. The pulmonary trunk leaves the right ventricle, pulmonary veins enter the left atrium, and the ascending aorta leaves the left ventricle. Look again for the pair of coronary arteries branching from the base of the aorta.

Working from the dorsal surface of the heart with a pair of fine scissors, cut through the caudal vena cava, through the wall of the right atrium, and into the right cranial vena cava (Incision 1, Fig. 4-5). Carefully pull out the coagulated blue injection mass. You may have to cut through the mass where it enters the right ventricle. The extent of the right atrium can now be seen. Identify the pulmonary veins returning from the lungs (Fig. 4-5) to the left atrium, cut through them, and extend your cut into the left auricle (Incision 2, Fig. 4-5). Open the left atrium and wash and pick out the dried blood. A thin **interatrial septum** separates the atria on the dorsal surface of the heart. The atria have thin walls, for they function primarily to accumulate blood during periods of ventricular contraction. When the ventricles relax and expand, the reduced pressure within them draws in the blood from the atria through the atrioventricular opening in the floor of each atrium. The muscles in the atrial walls, and those that crisscross the auricles, help only slightly in expelling the last of the blood into the ventricles.

Cut off the caudal one-third, or **apex,** of the heart in the transverse plane, and notice that the right and left ventricular cavities are separated by an **interventricular septum** (Fig. 4-6A). The left ventricle has a much thicker muscular wall than the right one does. It must develop sufficient pressure to pump blood to the entire body, except for the lungs. The thinner muscular walls of the right ventricle develop a pressure sufficient only to drive blood through the lungs and back to the heart. Too high a pulmonary pressure would cause an excessive filtration of fluid from the blood into the lungs.

Extend a diagonal incision (Fig. 4-6A, Incision 1) through the ventral wall of the right ventricle from the

Root of the Lung and Internal Structure of the Heart

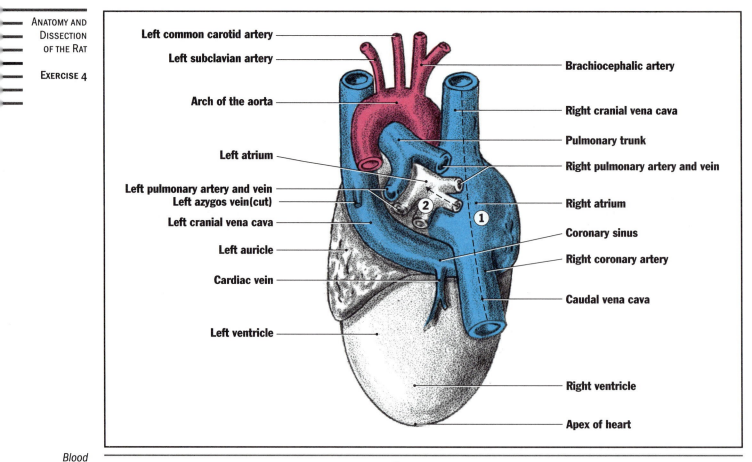

Left common carotid artery

Left subclavian artery

Arch of the aorta

Left atrium

Left pulmonary artery and vein
Left azygos vein(cut)

Left cranial vena cava

Left auricle

Cardiac vein

Left ventricle

Brachiocephalic artery

Right cranial vena cava

Pulmonary trunk

Right pulmonary artery and vein

Right atrium

Coronary sinus

Right coronary artery

Caudal vena cava

Right ventricle

Apex of heart

FIGURE 4-5
Dorsal view of the heart of a rat. The numbered dashed lines show
where the two incisions should be made to open the atria.

Blood

cut surface at the heart's apex into the pulmonary trunk. Remove the injection mass from the ventricle, but not from the pulmonary trunk. Try to avoid injuring the thin flaps that extend into the ventricle from the atrioventricular opening. There are three flaps that constitute the **right atrioventricular** (tricuspid) **valve,** but you may not see all of them. Each is held by delicate **tendinous cords** to small **papillary muscles** on the inner wall of the ventricle in such a way that blood easily enters from the atrium (Fig. 4-6B). When the ventricle contracts, the increase in pressure pushes the flaps together and prevents blood from returning to the atrium. Notice that the muscles next to the ventricular cavity form irregular ridges.

The injection mass in the pulmonary trunk is probably separated from that which was in the ventricle by three pocket-shaped flaps, the **pulmonary** (semilunar) **valve,** at the base of the trunk. It is sometimes easier to see the three bulges on the injection mass that are caused by this valve than the valve itself. This valve prevents blood in the pulmonary trunk from backing up during ventricular relaxation.

Open the left ventricle by extending an incision through its lateral wall from the apex to the left atrium (Fig. 4-6A, Incision 2), and clean out its contents. The **left atrioventricular** (bicuspid) **valve** is similar to the right one except that it has only two flaps. Look through the ventricle at the origin of the aorta and notice the **aortic valve.**

F. BLOOD

The blood vessels that you have been dissecting carry blood between different parts of the body. They are important pathways, but it is the blood that performs the transport, defense, and homeostatic functions of the circulatory system. Blood consists of a liquid **plasma** and several kinds of blood cells carried in the plasma. If blood smears from human beings or other mammals are available, or if you have the materials and stain needed to prepare them, the major types of blood cells can be examined.

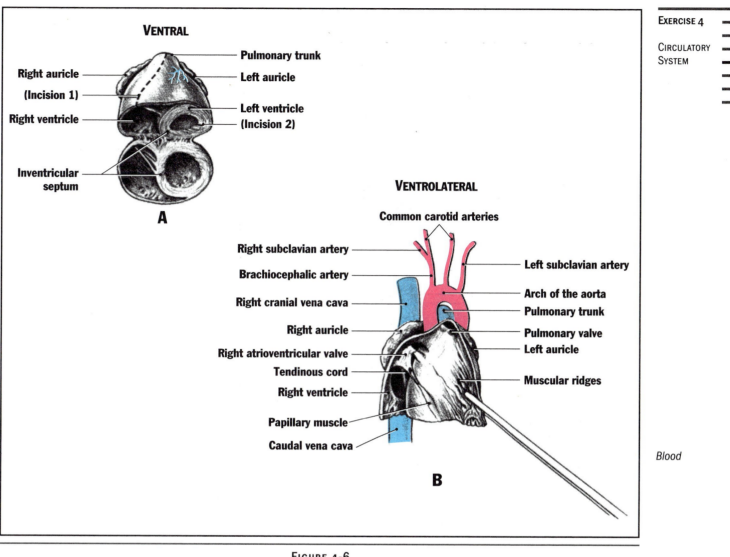

Blood

FIGURE 4-6
Dissection of the heart. (A) Caudoventral view after the apex has been cut off (the ventral surface of the heart is toward the top of the figure). A dashed line shows the incision to be made to open the ventricles. (B) Ventrolateral view after the right ventricle has been opened.

F.1. Erythrocytes

Red blood cells, or **erythrocytes,** are the most abundant type of blood cells, numbering 4.1 to 6.0 million per microliter of blood in an adult man and 3.9 to 5.5 million per microliter of blood in an adult woman. They appear as pinkish, circular cells about 8 micrometers in diameter (Fig. 4-7). Their nuclei, mitochondria, and most other cell organelles are lost in the course of their maturation in mammals, so they are shaped like biconcave disks. This is most evident in an edge view, but even in a surface view, the fact that the center of the cell is thinner than the periphery is usually discernible. This shape gives the cell a larger surface area relative to its volume than if it were a sphere.

Erythrocytes are filled with the respiratory pigment **hemoglobin,** which binds reversibly with oxygen, taking oxygen up in the lungs and releasing most of it in the tissues. Most of the carbon dioxide released by the tissues is carried in the plasma as carbonic acid or its ions and salts, but some also combines with hemoglobin and its transported to the lungs for release. Because erythrocytes lack nuclei and most cell organelles, they live only a short time, about 100 to 120 days. Cells lining the spleen and liver sinusoids capture and digest (**phagocytose**) senescent red blood cells. Phagocytosed cells are replaced by new ones that are continuously produced through mitosis from nucleated stem cells in the red bone marrow.

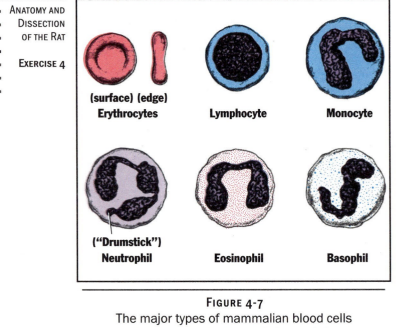

(surface) (edge)
Erythrocytes   **Lymphocyte**   **Monocyte**

("Drumstick")
Neutrophil   **Eosinophil**   **Basophil**

FIGURE 4-7
The major types of mammalian blood cells
can be distinguished by cell size and shape,
nuclear shape, and cytoplasmic granulation.

F.2. Leukocytes

Blood

The other blood cells are in life colorless or white and are referred to collectively as **leukocytes.** There are several kinds of leukocytes that number in aggregate about 5,000 to 9,000 per microliter of human blood. All leukocytes are nucleated, and most are larger than the erythrocytes (Fig. 4-7). Some have diameters as large as 20 micrometers. For the purpose of description, it is convenient to divide them into two groups: those having many conspicuous cytoplasmic granules (**granular leukocytes**) and those having very few cytoplasmic granules or none at all (**agranular leukocytes**). The granules have distinctive colors when treated with Wright's stain.

All leukocytes are capable of amoeboid movement, or, in other words, of squeezing between the endothelial cells of capillary walls, and of moving around within the tissues. They aggregate in areas of infection and inflammation. The relative numbers of the different types of leukocytes present in the blood are frequently diagnostic of specific diseases.

Lymphocytes are the most common agranular leukocytes representing from 20 to 30 percent of the leukocyte population. Most are slightly larger than erythrocytes and have a large, nearly spherical nucleus surrounded by only a thin layer of cytoplasm. Circulating lymphocytes contain few cytoplasmic organelles and are in a "resting" state, but they can change into active cells that play a critical role in the body's defense and in the development of immunity. B-lymphocytes can change into cells that produce antigens against invading antibodies. T-lymphocytes attack invading antigens directly in a cell-mediated reaction.

Monocytes are much larger and less common agranular leukocytes that make up from 2 to 8 percent of the white blood cells. The nucleus of a monocyte is typically kidney-shaped, and its cytoplasm is abundant. Monocytes change into tissue **macrophages.** These large and versatile cells play a role in the maintenance of normal tissues, but they are also important in phagocytosis, the development of immunity, and the repair of injured tissues. Monocytes are essentially macrophages that are in transit from the bone marrow, where they are produced, to the tissues.

Neutrophils, which constitute from 60 to 70 percent of all leukocytes, are the most common granular leukocytes. The nucleus is elongated, twisted, and usually constricted into three lobes connected by thin chromatin threads. Occasionally, part of the nucleus of a neutrophil in a female individual is set off as a small appendage called a **"drumstick,"** which consists of the chromatin of the female sex chromosomes. The cytoplasm has a fine, barely perceptible granulation that stains a light purple. Neutrophils can phagocytose microbes within the bloodstream, and they also collect in large numbers at sites of injury or infection, phagocytosing bacteria and dead cells. Their granules release enzymes that digest ingested particles. Therefore, the granules decrease in number as ingested particles are broken down. Neutrophils are one of the body's first lines of defense.

Eosinophils make up from 2 to 5 percent of the leukocytes. The nucleus usually consists of two lobes connected by a chromatin thread. Many relatively large, reddish granules fill the cytoplasm. Products released by the granules can kill some parasitic larvae within the circulatory system. Eosinophils collect in connective tissues near a body surface, such as the dermis and the lining of the digestive and respiratory system. Within the tissues, they appear to phagocytose various inflammatory substances and help rid the body of inactive antigen-antibody complexes.

Basophils are the rarest of the leukocytes, making up only 0.5 to 2.0 percent of the leukocyte population. The elongated and twisted nucleus frequently has an S-like configuration. The cytoplasm contains bluish granules of variable size. The granules of basophils contain histamine and heparin, which are released into injured tissues. Histamine dilates capillaries and increases their permeability to other leukocytes. Heparin is an anticoagulant that keeps blood flowing to injured tissues.

Some lymphocytes live for years, providing the basis for immunological memory and the development of immunity. But most leukocytes live for only a few days, because they are continuously being lost through the mucous membranes of the digestive, respiratory, and urinary tracts. Lymphocytes multiply and are stored in lymph nodes and other lymphoid tissues. All other leukocytes are produced through mitosis from nucleated stem cells in the red bone marrow.

F.3. Blood Platelets

Look for small groups of granules surrounded by bits of cytoplasm that are scattered among the other cells. These are the **blood platelets,** which are simply small blobs of cytoplasm that have budded off certain giant cells (**megakaryocytes**) in the bone marrow. They help protect the body from excess blood loss at the site of an injury: first, by clumping and helping to plug an injured blood vessel and, second, by releasing phospholipids. Phospholipids from injured platelets and **thromboplastin** from injured tissues initiate a series of enzyme-mediated reactions that result in the transformation of the soluble plasma protein fibrinogen into fibrin threads that form a blood clot.

Blood

Urogenital System

MHE UROGENITAL SYSTEM consists of two systems, the excretory system and the reproductive system. These two systems serve very different functions, but they are structurally so closely associated with each other that they share certain ducts and passages. This close anatomical association is a result of their development from adjacent embryonic precursors. The excretory system consists mainly of the kidneys, which have a vital role in eliminating nitrogenous waste products of the protein metabolism of the body cells. They also help to maintain homeostasis and regulate the pH as well as the balance of water, salt ions, sugars, and many other substances in the body fluids. The reproductive system produces gametes for sexual reproduction and, in female mammals, serves also to protect and nourish the embryo until it is mature enough to be born.

A. EXCRETORY SYSTEM

You have already observed the pair of bean-shaped **kidneys** lying against the back muscles dorsal to the parietal peritoneum (Exercise 4). Each is drained by a slender duct, the **ureter,** which leaves from the slightly indented medial margin of the kidney, accompanied by the renal blood vessels. Trace a ureter caudally along the back muscles. Near the pelvic canal, it turns ventrally and follows the vesicular artery. It goes deep to the **ductus deferens** (sperm duct), which emerges from the scrotum in a male (see Fig. 5-3) or deep to the **horn of the uterus** in a female (see Fig. 5-5) to enter the base of the **urinary bladder.** Urine is temporarily stored in the urinary bladder, but this organ is usually contracted and empty in a preserved specimen and appears as a small, pear-shaped, muscular sac. It is attached to the body by a midventral mesentery and to the brim of the pelvic canal by a pair of lateral mesenteries, which often contain wads of fat and, in the male, part of the prostate (see later). Carefully tear these mesenteries, push the bladder dorsally, and trace it toward the pelvic canal. Its base, or neck, narrows to become a **urethra,** which continues through the pelvic canal. You will follow the urethra later in this exercise when you study the reproductive passages.

Excretory System

Dissect away connective tissue and fat from around the renal vessels and ureter of one kidney and cut off its ventral half or, in other words, section the kidney in the frontal plane of the body (Fig. 5-1). You do not need to remove the kidney to perform this section; the dorsal half of the kidney can be left in its original place. As the ureter enters the kidney, it expands to form a funnel-shaped cavity, the **renal pelvis.** The renal blood vessels branch near the kidney and enter the substance of the kidney beside the ureter. The kidney tissue consists of a peripheral, lighter, somewhat granular **renal cortex** and a central, darker, somewhat striated **renal medulla.** The renal medulla converges to form a conspicuous **renal papilla** which projects into the renal pelvis.

The structural and functional unit of the kidney is the **nephron.** It is comprised of the **glomerulus,** a capillary tuft, and the **renal tubule,** which receives fluid from the glomerulus and processes it to become urine. The various parts of the nephron have specific functions and are located in specific places within the cortex and medulla of the kidney (see Fig. 5-2). Blood coming from the **renal artery** (see Exercise 4) at a slightly higher pressure than in other arteries of the body is distributed to **afferent arterioles** within the kidney. Blood in each afferent arteriole enters the capillaries of the glomerulus where it is filtrated into the cavity of the **glomerular capsule** (of Bowman). Together, the glomerulus and the glomerular capsule form a structural and functional unit

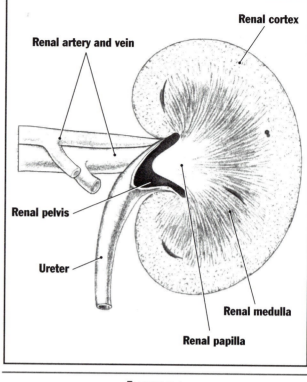

FIGURE 5-1
Longitudinal section of the kidney of a rat.

called a **renal corpuscle.** The glomerular filtrate in the glomerular capsule is very similar in composition to blood, except that it does not contain blood cells and larger protein molecules, which are too large to pass through the epithelial layer of the capillaries and glomerular capsule. The glomerular filtrate continues on through the **proximal convoluted tubule** in the renal cortex, through the **loop of Henle** into and out of the renal medulla, through the **distal convoluted tubule** back in the renal cortex, and finally enters the **collecting tubule,** which extends from the renal cortex through the renal medulla to the tip of a renal pyramid. As the glomerular filtrate passes through the various parts of a renal tubule, water, ions, nutrients, and other substances needed by the body are absorbed by the walls of the renal tubule and returned to the blood stream through the **peritubular capillaries.** The peritubular capillaries are supplied by the **efferent arterioles,** which collect blood from the glomerulus, and are drained by **renal venules,** which are collected by the **renal vein.** Waste products, in particular ammonia and urea, remain in the glomerular filtrate. In this manner, the metabolic waste products become more and more concentrated in the glomerular filtrate, which is transformed into urine by the time it leaves the renal pyramid to enter the ureter.

B. REPRODUCTIVE SYSTEM

The reproductive system produces and transports the gametes for sexual reproduction and, in females, serves also to protect and nourish the developing embryo. Parts of the testes and ovaries function as endocrine organs that produce hormones and regulate many aspects of growth and reproduction. Although you will dissect the reproductive system of only one sex, you should use another student's specimen to study the opposite sex. Students should therefore be particularly careful in dissecting their own specimens and should prepare dissections that will be clear enough for others to study.

B.1. Male Reproductive Organs

Insert a pair of scissors into the preputial orifice and cut open the **prepuce,** or foreskin, which is the sheath of skin that covers the tip of the penis. Look for a pair of nearby subcutaneous **preputial glands** that discharge an

odoriferous secretion, the **smegma,** into the **preputial pouch.** Uncover more of the penis so that it will not be inadvertently cut in later stages of this dissection and then remove the skin layer of the scrotal sac. The scrotum will vary greatly in size among different specimens. During the breeding season, it is large because the testes lie within it, but the testes are withdrawn into the body cavity during nonreproductive periods in rodents and some other mammals. Sperm cells of most mammals cannot fully mature at the high temperatures occurring in the abdomen; the slightly lower temperature in the scrotum is necessary for the final stages of sperm formation.

The **scrotum** is essentially a cutaneous sac that contains paired extensions of the abdominal cavity in which the testes are housed (Fig. 5-3). Very carefully remove the skin and the underlying connective tissue; you will discover the paired **cremasteric pouches** with the testes inside them. The wall of a cremasteric pouch consists of connective tissue, which is an extension of the fasciae of the body wall, and of the **cremasteric muscle,**

*Reproductive
System*

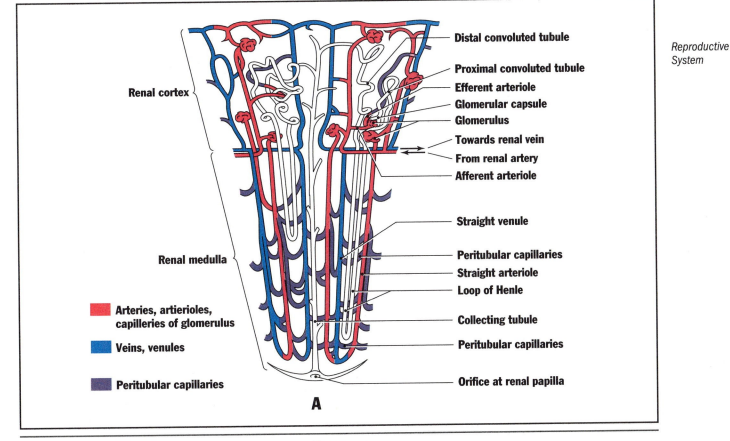

A

- Arteries, artierioles, capilleries of glomerulus
- Veins, venules
- Peritubular capillaries

Labels:
- Distal convoluted tubule
- Proximal convoluted tubule
- Efferent arteriole
- Glomerular capsule
- Glomerulus
- Towards renal vein
- From renal artery
- Afferent arteriole
- Straight venule
- Peritubular capillaries
- Straight arteriole
- Loop of Henle
- Collecting tubule
- Peritubular capillaries
- Orifice at renal papilla
- Renal cortex
- Renal medulla

FIGURE 5-2
Diagrams illustrating the arrangement of nephrons and blood vessels in a mammalian kidney. (A) Two adjacent nephrons and several glomeruli and blood vessels.

ANATOMY AND
DISSECTION
OF THE RAT

EXERCISE 5

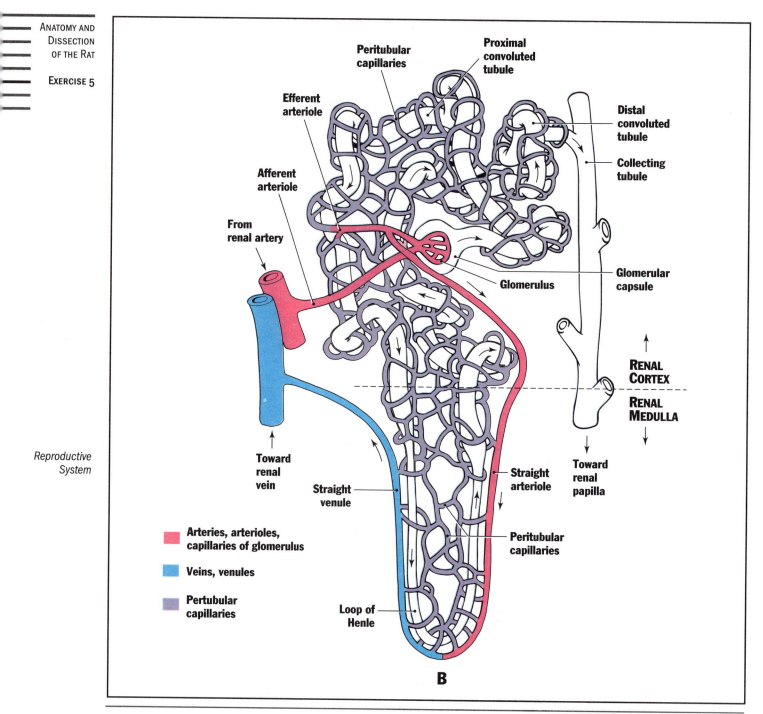

FIGURE 5-2 CONTINUED
Diagrams illustrating the arrangement of nephrons and blood vessels in a mammalian kidney. (B) Enlargement of a single nephron with associated capillaries. (Redrawn from Walker, W. F., Jr., and Homberger, D. G. *Vertebrate Dissection,* 8th ed. Philadelphia: Saunders College Publishing, 1992. Dorit, R., Walker, W. F., Jr., and Barnes, R. D. *Zoology.* Philadelphia: Saunders College Publishing, 1991. Stalheim-Smith, A. and Fitch, G. K. *Understanding Human Anatomy and Physiology.* Minneapolis/St. Paul: West Publishing Company, 1993. Dyce, K. M., Sack, W. O., and Wensing, C. J. G. *Textbook of Veterinary Anatomy.* Philadelphia: W. B. Saunders Company, 1987.)

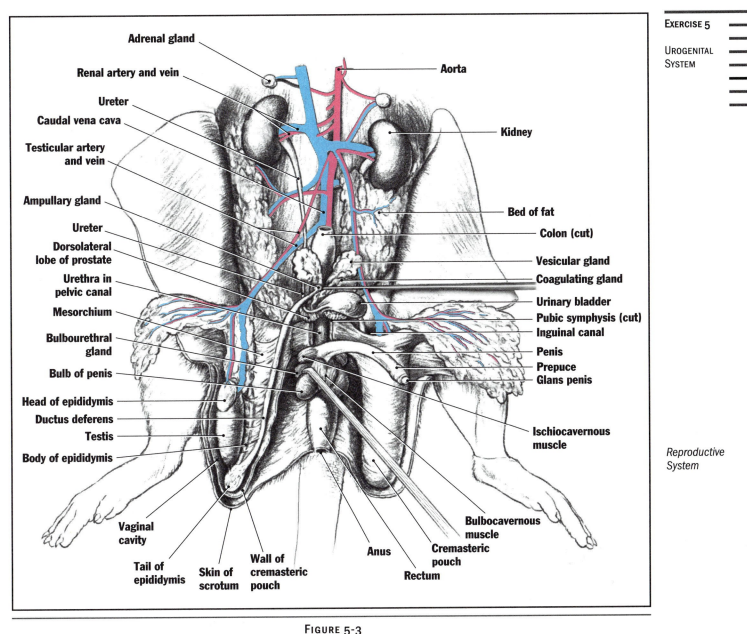

Adrenal gland

Renal artery and vein

Ureter

Caudal vena cava

Testicular artery
and vein

Ampullary gland

Ureter

Dorsolateral
lobe of prostate

Urethra in
pelvic canal

Mesorchium

Bulbourethral
gland

Bulb of penis

Head of epididymis

Ductus deferens

Testis

Body of epididymis

Aorta

Kidney

Bed of fat

Colon (cut)

Vesicular gland

Coagulating gland

Urinary bladder

Pubic symphysis (cut)

Inguinal canal

Penis

Prepuce

Glans penis

Ischiocavernous
muscle

Vaginal
cavity

Tail of
epididymis

Skin of
scrotum

Wall of
cremasteric
pouch

Anus

Rectum

Cremasteric
pouch

Bulbocavernous
muscle

FIGURE 5-3
Ventral view of the urogenital system of a male rat. The pelvic canal has been opened.

which is an extension of the internal oblique and transversus abdominis muscles of the abdominal wall (Exercise 2). The cremasteric muscle is well developed in the rat because it is used to retract the testes into the peritoneal cavity. In the male human being, it is atrophied.

The cavity within the cremasteric pouches is a caudal extension of the peritoneal cavity and is called the **vaginal cavity.** To reach the cremasteric pouch, the vaginal cavity has to pass through the muscular abdominal wall; this passage through the body wall is called the **inguinal canal.**

Leave one cremasteric pouch intact so that you can demonstrate it to a classmate who is dissecting a female specimen. Cut open the other one along its ventral sur-

face. The vaginal cavity is in broad communication with the peritoneal cavity to permit the passage of the testes when they are pulled back into the peritoneal cavity. In a male human being, the testes lie permanently within the scrotum, and the proximal part of the vaginal cavity near the body wall has completely shrunk to let only the sperm duct, nerves, and blood vessels pass through. Because the vaginal cavity is an extension of the coelom, it is lined with serosa, called the **parietal tunica vaginalis;** and the organs lying within are enveloped by serosa, called the **visceral tunica vaginalis** (Fig. 5-4A). A dorsal mesentery, the **mesorchium,** extends from the wall of the vaginal cavity to the organs. At its very caudal end, the mesorchium is thickened somewhat to form the

69

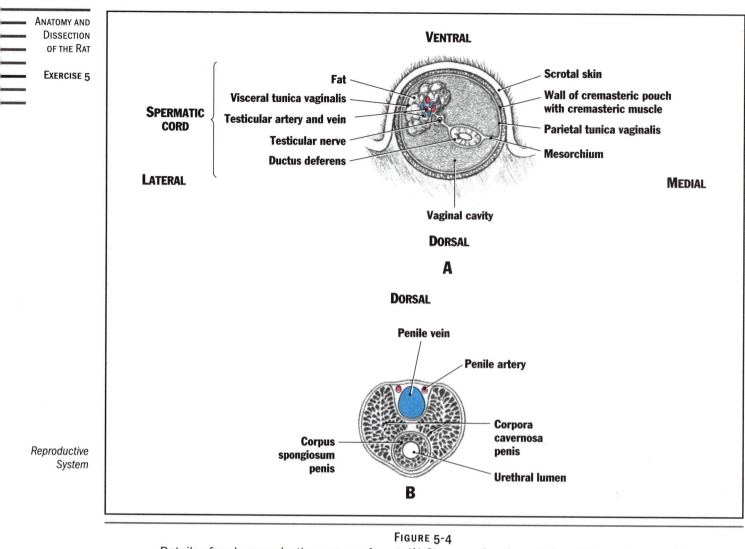

VENTRAL

SPERMATIC
CORD

LATERAL

Fat
Visceral tunica vaginalis
Testicular artery and vein
Testicular nerve
Ductus deferens

Scrotal skin
Wall of cremasteric pouch
with cremasteric muscle
Parietal tunica vaginalis
Mesorchium

MEDIAL

Vaginal cavity

DORSAL

A

DORSAL

Penile vein

Penile artery

Corpus
spongiosum
penis

Corpora
cavernosa
penis

Urethral lumen

B

FIGURE 5-4
Details of male reproductive organs of a rat. (A) Cross section through the right scrotum cranial
to the testis (the dorsal surface is toward the bottom of the figure). (B) Cross section
through the middle of the penis (the dorsal surface is toward the top of the figure).

gubernaculum. The early development of the gubernaculum in the fetus and its failure to lengthen as much as the rest of the body cause the initial descent of the testes from the abdominal cavity.

The **testis** is an oval organ lying near the caudal end of the vaginal cavity (Fig. 5-3). In addition to producing the sperm cells, it secretes the hormone testosterone, which stimulates the production of sperm cells and the development of male secondary sex characteristics. Dissect away the fat at the cranial end of the testis and trace the testicular blood vessels to it; the testicular artery is highly convoluted and the accompanying testicular vein becomes a network of intertwining vessels known as the **pampiniform plexus.** The proximity of the testicular artery to the veins of the pampiniform plexus enables heat to flow from the testicular artery,

which is filled with warm blood coming from the body core, to the veins of the pampiniform plexus, which is filled with cooler blood returning from the peripheral testis. In this manner, the oxygen-rich blood reaching the testis is already somewhat cooled.

A crescent-shaped **epididymis** partly surrounds the testis. Sperm cells produced in seminiferous tubules leave the testis through microscopic tubules that form a network, the **rete testis,** in the cranial end of the testis. In the rat, part of the rete testis forms a cordlike structure that leads from the testis to the **head of the epididymis** at the cranial end of the testis. It can be found by dissecting between the testis and the head of the epididymis. Sperm cells then continue through the **body of the epididymis** to its **tail,** where they enter the **ductus deferens,** or sperm duct. Look at the epididymis carefully and notice that much of it is composed of a highly coiled tube in which sperm cells mature and are stored. Trace the ductus deferens cranially to the body cavity. It accompanies the testicular blood vessels; together, the testicular blood vessels and nerves, the ductus deferens, and the enveloping visceral tunica vaginalis constitute the **spermatic cord** (Fig. 5-4A).

Pull the two cremasteric pouches laterally and trace the penis across the ventral wall of the pelvis to the caudal end of the pelvic canal. Pull the penis caudally and cut away midventral abdominal muscle to expose the pubic symphysis. Cut through the pubic symphysis and pull the hind legs laterally and dorsally to spread open the pelvic canal.

Follow the ductus deferens into the abdominal cavity. It loops over the ureter and turns caudally between the urinary bladder and a group of accessory genital glands to join the urethra. The most ventral accessory genital glands are the paired **ventral lobes** of the **prostate,** which lie in the mesentery on each side of the urinary bladder. Push them aside so that you can trace the ureter to its union with the urethra (Fig. 5-3). As you search for this union, cut the testicular blood vessels as well as the vesicular blood vessels supplying the urinary bladder on one side so that you can approach the urethra laterally and dorsally as well as from its ventral surface. The largest of the group of accessory genital glands is a pair of **vesicular glands,** which can be recognized by their serrated lateral borders and narrow hooked ends. Paired **coagulating glands,** which are quite large, lie close to the ventral surface of the vesicular glands; they should be separated by breaking the connective tissue binding them together. The vesicular and coagulating glands join the urethra dorsally beside the entrances of the paired ductus deferentes. The terminal part of each ductus deferens is slightly swollen to form a small **ampullary gland** (gland of the ductus deferens). The finely lobulated **dorsolateral lobe** of the **prostate** lies dorsal to the base of the vesicular glands and the urethra, but part of it swings around onto the ventral surface of the urethra and nearly obscures the entrance of the ductus deferens and other glands. Many small ducts from all parts of the prostate enter the urethra directly, both dorsally and ventrally. The secretions of the ampullary glands and the prostate form the **seminal fluid,** which carries the sperm cells during ejaculation, activates and provides certain nutrients for them, and contains substances that neutralizes the slightly acidic environment in the vagina. The vesicular and coagulating glands of the rat release their secretions only toward the end of ejaculation. These two secretions mix, coagulate in the vagina, and form a **vaginal plug.** This vaginal plug prevents sperm cells from subsequent copulations with possibly different males from entering the uterus and oviducts and from fertilizing additional mature ova.

Trace the urethra caudally, separating it from the underlying **rectum.** It enters the penis at the outlet of the pelvic canal. The base of the **penis** is held in place by a pair of **ischiocavernous muscles** which extend laterally and dorsally to attach to the pelvic girdle. Cut through one of the muscles and notice that it surrounds a column of spongy, vascular erectile tissue, the **corpus cavernosum penis,** which joins its counterpart from the other side of the body and extends along the dorsal surface of the penis (Fig. 5-4B). The base of the penis is swollen caudal to the ischiocavernous muscles and is covered by the **bulbocavernous muscle,** part of which passes dorsally, lying lateral to the rectum. Strip off this sheet of muscle and notice the paired **bulbs of the penis.** Cut into one of them and observe that it, too, is filled with erectile tissue. This tissue is a part of the **corpus spongiosum penis,** which surrounds the urethra and extends distally on the ventral surface of the penis. Another pair of accessory genital glands, the **bulbourethral glands,** lie dorsal to the bulbs of the penis.

Make a cross section near the middle of the penis and observe the extensions of the paired corpora cavernosa penis and of the corpus spongiosum penis. You will also see penile arteries and veins (Fig. 5-4B). An erection of the penis is caused when the vascular spaces within the erectile tissue become engorged with blood. The part of the penis distal to the attachment of the prepuce is the **glans penis,** which is part of the corpus spongiosum penis. In rodents and many other mammals, it is covered by a horny and spiny epithelium. Cut across the glans and notice that the tip of the penis is stiffened by a small **penis bone** lying dorsal to the urethra.

B.2. Female Reproductive Organs

Notice again the external orifice of the urethra, which is in a small depression called the **fossa clitoridis,** and the orifice of the vagina in a larger depression called the **vulva.** Extend the longitudinal abdominal incision, made when you opened the body cavity, caudally through the skin on one side of these orifices to the anus if you have not already done so, and then cut through the side of the fossa clitoridis and intersect the previous incision. As you do so, notice the pair of subcutaneous **glands of the clitoris** (Fig. 5-5). Spread open the fossa clitoridis and notice the pointed tip of the clitoris (**glans clitoridis**) lying just ventral to the urethral opening. Trace the clitoris caudally and dorsally toward the outlet of the pelvic canal. This part of the clitoris is composed of paired columns of erectile tissue (the **corpora cavernosa clitoridis**) that are connected by ligaments to the pelvic girdle. All these structures are comparable to structures in a male: the wall of the fossa clitoridis corresponds to the prepuce; the glands of the clitoris to the preputial glands; the glans clitoridis to the glans penis; the corpora cavernosa clitoridis to the corpora cavernosa penis. Notice their resemblance when you compare male and female specimens.

Reproductive System

ANATOMY AND
DISSECTION
OF THE RAT

EXERCISE 5

Reproductive
System

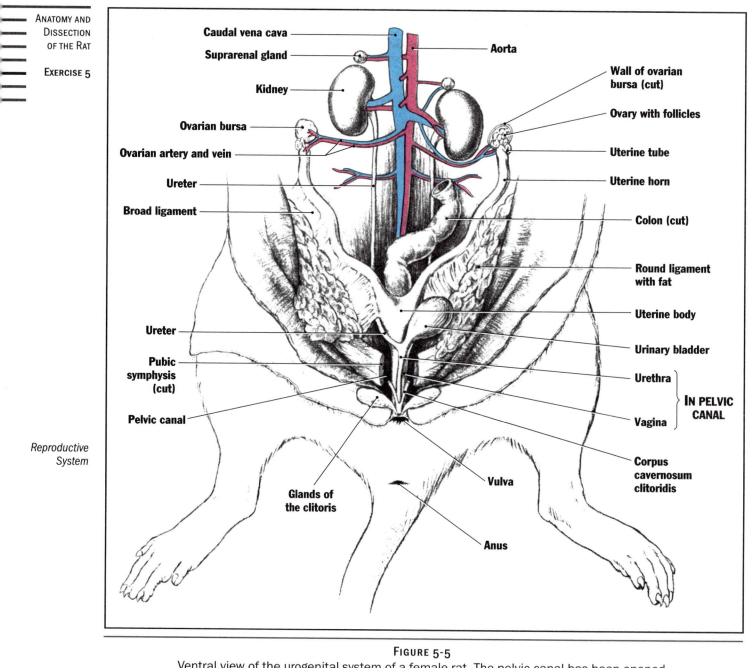

FIGURE 5-5
Ventral view of the urogenital system of a female rat. The pelvic canal has been opened.

Cut away midventral abdominal muscles to expose the pubic symphysis; then cut through the pubic symphysis, pull the hind legs of the animal laterally and dorsally, and spread open the pelvic canal. Find again the urinary bladder, and the ureters entering its base. The urinary bladder narrows to become the **urethra,** which can now be traced caudally through the pelvic canal to its orifice dorsal to the clitoris. Remove the connective tissue, fat, and blood vessels from the urethra and from the distal part of the reproductive tract, the **vagina,** that lies dorsal to the urethra.

The **uterus** is a Y-shaped organ. It consists of a pair of elongate, independent **uterine horns.** Caudally, the uterine horns are surrounded by a common outer layer of musculature and form the short **uterine body** just before opening separately into the vagina. You can see the openings of the uterine horns if you push the bladder and urethra aside, make a midventral incision along the length of the vagina, and spread open the vaginal walls (Fig. 5-6). The part of the uterus that bears the openings of the uterine horns protrudes a short distance into the vagina and is known as the

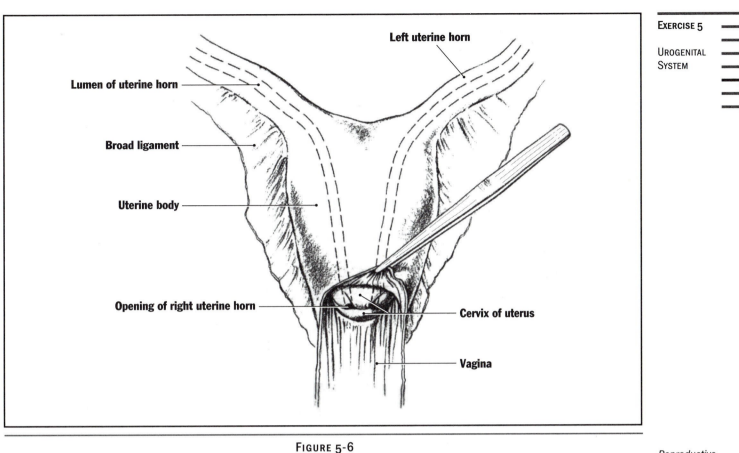

FIGURE 5-6
Ventral view of the uterus and vagina of a female rat.

neck, or **cervix,** of the uterus. The degree to which the uterine horns remain independent units varies considerably among mammals. For example, in rabbits, the uterine horns remain completely independent, and each side has its own cervix. As we have seen in rats, the uterine horns are independent but open within the same cervix, though separately. In carnivores, such as cats, the uterine horns are separate, but open through a common opening in the cervix. In human beings, the uterine horns have fused to form a single uterus with a single opening in the cervix.

Each uterine horn is anchored to the dorsal body wall by a mesentery known as the **broad ligament** (Fig. 5-5). Another mesentery is attached to the broad ligament and runs perpendicularly across its ventral surface into a recess of the body cavity that extends a short distance caudally ventral to the pelvic girdle. This mesentery contains a great deal of fat in the rat, but in many mammals it is cordlike and therefore called the **round ligament of the uterus.** It corresponds to the gubernaculum in a male specimen. The abdominal recess into which the round ligament of the uterus extends corresponds to the vaginal cavity in a male specimen.

Trace a uterine horn cranially. Near the caudal end of the kidney, it narrows abruptly into a very small and highly convoluted **uterine tube** (Fallopian tube). In many mammals, the cranial end of the uterine tube expands to form a funnel-shaped infundibulum, which receives the eggs from the adjacent ovary; in rats and mice, however, the infundibulum and adjacent parts of the broad ligament form a membranous sac, the **ovarian bursa,** which completely envelops the ovary.

Pick away fat at the cranial end of the uterine tube, cut open the ovarian bursa, and study the small ovary within. The ovary is enveloped by serosa, and a mesentery anchors it to the wall of the ovarian bursa, which in turn is anchored to the body wall by a mesentery. In a sexually mature female specimen, many small liquid-filled vesicles protrude from the surface of the ovary; each vesicle is a follicle containing a maturing egg. At ovulation, several greatly swollen follicles rupture, breaking through the thinned ovarian wall and serosa, and release their eggs into the ovarian bursa. The cells of the follicles that remain in the ovary grow into the **corpora lutea** (see Section C.2). Corpora lutea are maintained only if there is a pregnancy; they atrophy at the end of the pregnancy.

Follicles and corpora lutea are endocrine glands: follicles secrete the hormone estrogen, and corpora lutea produce the hormone progesterone. Both hormones are involved in regulating the reproductive cycles.

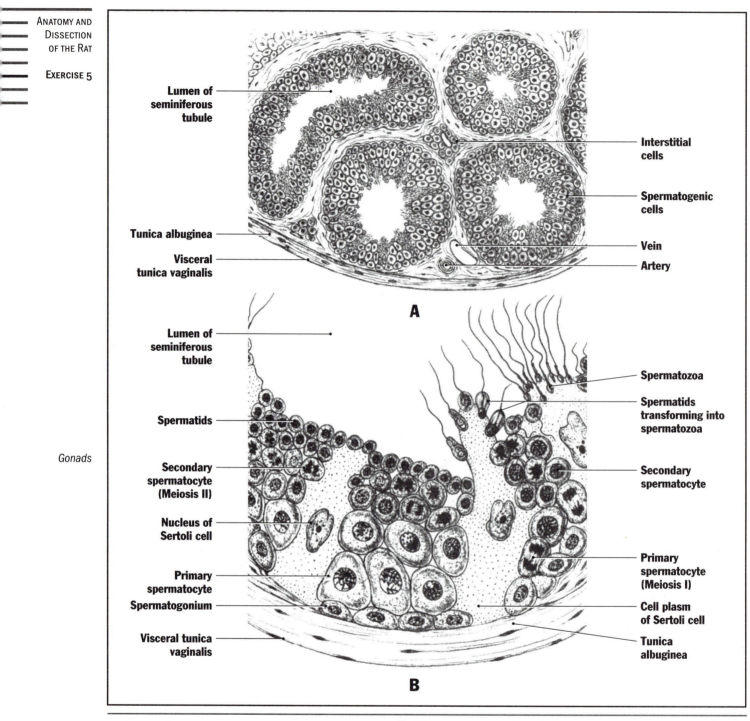

FIGURE 5-7
Diagram of a cross section of a histological slide preparation of a part of a mammalian testis. (A) Low magnification of several seminiferous tubules. (B) High magnification of part of one seminiferous tubule.

C. GONADS

The gonads—namely, the testes and ovaries—produce gametes, the sperm cells and eggs. To study gamete production, observe histological slide preparations of mammalian gonads.

C.1. Testis

The **testis** is composed primarily of many long and highly coiled **seminiferous tubules,** of which some will be seen cut in cross section and some in oblique section (Figs. 5-7 and 5-8). The total length of all of them in a human testis is estimated to be 250 meters! A tough,

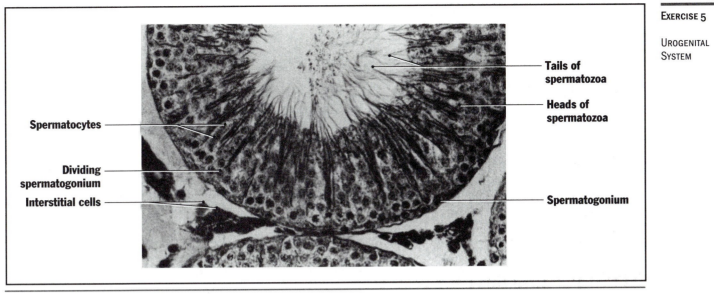

Spermatocytes

Dividing
spermatogonium

Interstitial cells

Tails of
spermatozoa

Heads of
spermatozoa

Spermatogonium

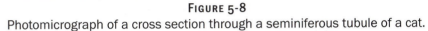

FIGURE 5-8
Photomicrograph of a cross section through a seminiferous tubule of a cat.

fibrous capsule, the **tunica albuginea,** forms the wall of the testis and sends septa into the organ. A thin serosa, the visceral tunica vaginalis, envelops the testis.

Examine the walls of several seminiferous tubules. Chromosomes will be visible in many of the cells, for they are dividing rapidly by mitosis. The wall of the seminiferous tubule is occupied by diploid spermatogonia, which start to divide mitotically at puberty. Some of their daughter cells remain near the wall of the seminiferous tubule as spermatogonia; others move toward the open center, or lumen, of the seminiferous tubule and become **primary spermatocytes.** Spermatogonia and primary spermatocytes contain the diploid number of chromosomes. Each primary spermatocyte undergoes a meiotic division to form two **secondary spermatocytes,** and each of these divides meiotically again to form two **spermatids.** Thus, four haploid spermatids develop from a single diploid primary spermatocyte.

Cells in the testis do not grow between the successive cell divisions, so they become smaller and smaller as they move closer to the center of the seminiferous tubule. The various stages of spermatogenesis are most easily identifiable by the relative size and the relative position of the cells within the seminiferous tubules.

A spermatid undergoes a transformation into a mature, motile **spermatozoon.** Its nucleus condenses and appears very small and dark, much of the cytoplasm is lost, and a tail develops. During this process, it becomes more evident that the spermatogenic cells are closely associated with large **Sertoli cells,** which extend from the wall to almost the center of a seminiferous tubule between groups of spermatogenic cells. The cell membranes of the Sertoli cells are difficult to see with ordinary stains, but their cell

bodies can be discerned by observing the groups of maturing sperm cells surrounding them.

Groups of relatively large, light-staining **interstitial cells** are present in the connective tissue between the seminiferous tubules. They produce **testosterone,** the hormone responsible for the production of sperm cells and the development of male secondary sexual characteristics. Spermatogenesis is cyclical in seasonally breeding mammals, such as the rat, but continuous in males of human beings. Separate groups of spermatogenic cells mature at different times; so the particular combination of types of spermatogenic cells may differ in different tubules or in different parts of the same tubule. Some groups of spermatogenic cells may be in the early stages of spermatogenesis; others, in the spermatid or spermatozoa stages. It probably takes 16 days for human spermatozoa to develop from spermatogonia, but the area of sperm production is so large that vast numbers of sperm cells are produced, as many as 200 to 300 million per ejaculation.

C.2. Ovary

In contrast to the testis, which consists essentially of a mass of tubules, the ovary consists of a solid mass of cells and tissues. A small **ovarian medulla** in the center of the ovary consists of highly vascularized connective tissue. The **ovarian cortex** consists of dense connective tissue containing follicles with maturing eggs (Figs. 5-9 and 5-10). The cortical connective tissue forms a very condensed layer around the ovary, which is devoid of follicles and is called **tunica albuginea.** The surface of the ovary is covered by a layer of serosa, or visceral peri-

Gonads

75

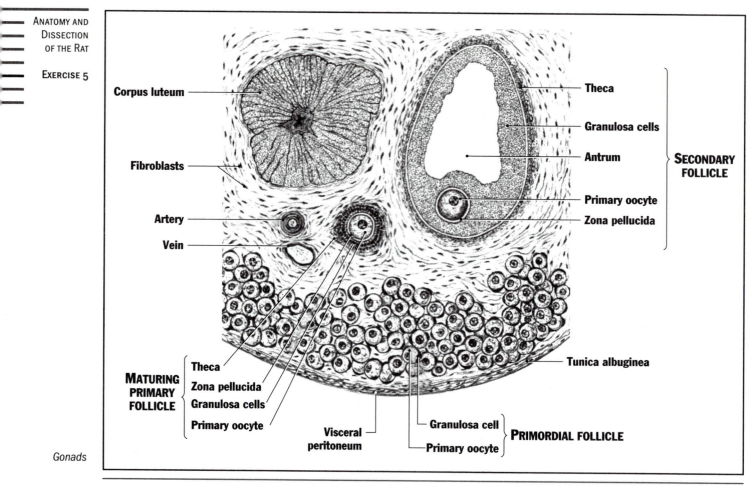

FIGURE 5-9
Diagram of a section of a histological slide preparation of a segment of the ovarian cortex of a mammal.

toneum. The periphery of the ovarian cortex just below the tunica albuginea is filled with several layers of **primordial follicles.** Each primordial follicle contains a large **primary oocyte** characterized by a conspicuous eccentric nucleus embedded in a relatively clear cytoplasm. A single layer of flattened follicle cells, known as **granulosa cells,** surrounds each primary oocyte. You will also see one or more maturing **primary follicles** in which the primary oocyte is surrounded by a light-staining region known as the **zona pellucida** and by one or several layers of granulosa cells that have assumed a cuboidal shape. You will also discover **secondary follicles** in which a liquid-filled space, the **antrum,** has developed and is surrounded by the greatly increased number of granulosa cells. Finally, you may be able to find a **corpus luteum,** which consists of granulosa cells that have proliferated after the mature egg had left the mature **tertiary follicle,** also called **Graafian follicle,** at ovulation.

Unlike gamete development in the testis, where the maturation of spermatogonia into mature sperm cells is a continuous process and can be reconstructed by observing a single histological slide preparation, the maturation

from oogonia to mature eggs (**ova**) cannot be observed in any one histological slide preparation of an ovary. All the stages of ovum development are not visible in one section because this maturation process takes place in a stepwise fashion at different stages in the life of a female mammal.

During embryonic development the diploid oogonia multiply by mitosis and mature into diploid primary oocytes by starting the first meiotic division up to the prophase. They are surrounded by a layer of flattened granulosa cells and form primordial follicles. The primary oocytes stop dividing at birth, when there are about 200,000 to 500,000 of them per ovary of a female human being. This number is more than enough to last through the reproductive years of an individual, during which usually only one mature egg is released in each menstrual cycle, even though tens of thousands of primary oocytes atrophy without ever maturing to the stage of ovulation. The primordial follicles remain dormant from birth to puberty in human beings and until sexual maturity in the rest of the mammals.

Beginning with puberty or sexual maturity, the follicles and maturing eggs undergo recurring **ovarian**

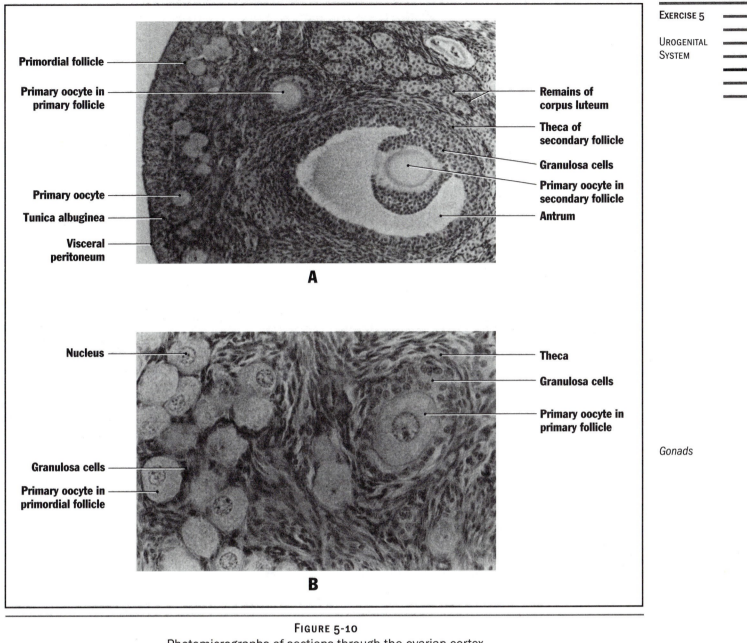

Primordial follicle

Primary oocyte in
primary follicle

Primary oocyte

Tunica albuginea

Visceral
peritoneum

Remains of
corpus luteum

Theca of
secondary follicle

Granulosa cells

Primary oocyte in
secondary follicle

Antrum

A

Nucleus

Granulosa cells

Primary oocyte in
primordial follicle

Theca

Granulosa cells

Primary oocyte in
primary follicle

B

FIGURE 5-10
Photomicrographs of sections through the ovarian cortex
of a cat. (A) Low magnification. (B) High magnification.

Gonads

cycles, during which the follicles enlarge and the eggs mature and are eventually discharged from the ovary. The number of eggs released and the duration and frequency of the ovarian cycle varies with the species. Deer and other seasonally breeding mammals go through one or more ovarian cycles during a brief reproductive season, and the ovary is dormant for most of the year. Cats, dogs, and many other domestic mammals undergo two or more reproductive periods per year. Many other mammals, including rats and human beings, have repetitive ovarian cycles throughout the year.

In the human being, the ovarian cycle lasts an average of 28 days, and is accompanied by a **menstrual cycle,** during which the uterine lining undergoes changes, such as growth and increased vascularization, in preparation for the implantation of a fertilized egg. If no fertilized egg is implanted into the uterine lining, the accrued uterine tissue is shed through a process known as menstruation, and a new reproductive cycle begins. In the rat, the reproductive cycle lasts only 4 to 5 days, and between 4 and 12 ova are ovulated per ovarian cycle. If no fertilized eggs are subsequently implanted in the uterine lining, the accrued uterine tissue regresses gradually without any bleeding. This lack of menstruation is also characteristic of most other mammals, with the exception of human beings and old-world primates.

Starting at puberty in human beings, several primary oocytes start to mature each month and acquire a zona pellucida. The granulosa cells assume a cuboidal shape and multiply by mitosis, thereby forming a multilayered primary follicle. The primary oocyte now completes meiosis I, generating a diploid **secondary oocyte** that has all the cytoplasm and a diploid **primary polar body** that has no cytoplasm and is destined to disintegrate. Concurrently, the granulosa cells continue to multiply, and a liquid-filled cavity, the antrum, develops in their midst. At this stage, the follicle is called a secondary follicle. The antrum continues to increase in size as more and more liquid accumulates in it.

Before ovulation, one of the maturing secondary follicles accelerates its growth and moves toward the surface of the ovary. In human beings, it is usually only this one follicle that will become a tertiary follicle and ovulate. At ovulation, this tertiary, or Graafian, follicle ruptures through the thinned tunica albuginea and serosa and releases the secondary oocyte into the ovarian bursa. As the secondary oocyte enters the uterine tube, it starts to divide through meiosis II, but it remains in the metaphase stage until it is fertilized.

Gonads

At fertilization, which must take place in the uterine tube within about 24 to 72 hours after ovulation in human beings, the meiotic division of the secondary oocyte is completed by generating a mature haploid oocyte (ovum) that has all the cytoplasm and a haploid **secondary polar body** that has no cytoplasm and is destined to die.

After ovulation, the granulosa and thecal cells multiply, enlarge, and accumulate lipids and other substances. These cells form a corpus luteum, which lasts in

human beings until shortly before the onset of the next menstruation and then regresses. This regression marks the end of an ovarian cycle. If a fertilized egg implants and pregnancy ensues, however, the corpus luteum continues to grow and lasts well into the pregnancy, when the placenta takes over as the main hormone-producing organ during the later stages of pregnancy.

The follicles do not only protect and nourish the developing eggs, but they and the corpora lutea are also endocrine glands. The follicles develop under the influence of **follicle stimulating hormone** secreted by the anterior part of the hypophysis (Exercise 6). As the follicles mature, they themselves secrete **estrogen,** which is responsible for the enlargement of the uterus and the growth of the uterine lining following menstruation. The corpora lutea develop under the influence of another hypophyseal gonadotropic hormone, the **luteinizing hormone.** The corpora lutea also produce some estrogen, but they secrete primarily **progesterone.** This hormone is needed for the final development and increased vascularization of the uterine lining in preparation for the implantation of a fertilized egg, as well as for the maintenance of the uterine lining and placenta during the early stages of pregnancy.

As we have seen, sperm cells are very small, mainly because they possess very little cytoplasm, whereas ova possess all the cytoplasm that the original oogonium had and continued to accumulate as it matured into a primary oocyte. Consequently, the cytoplasm of the zygote resulting from the union of one sperm cell and one ovum originates almost entirely from the maternal ovum.

A fertilized ovum travels through the uterine tube in about 7 days in the female human being and implants in the lining of the uterus, where a placenta will develop from both maternal and fetal tissues.

EXERCISE

Six

6

Nervous Coordination: Nervous System

URVIVAL OF ANY ORGANISM requires the harmonious interaction of its various organs and appropriate responses of the entire organism to changes in the external and internal environments. Many aspects of metabolism, growth, and reproduction are regulated by the secretions of the numerous endocrine glands. (The thyroid gland, islets of Langerhans, and adrenal gland have been mentioned in previous exercises. The hypophysis and pineal body will be considered during the study of the brain.) The circulatory system carries minute quantities of hormones secreted by a gland in one part of the body to other parts of the body. Some hormones affect most cells in the body, but some cells have receptor molecules specific for particular hormones. Endocrine integration tends to be slow, widespread, and long lasting in its effects. Feedback mechanisms of various types control the activity of the endocrine glands and the hormonal levels in the blood.

More rapid and specific integration of the internal organs, as well as most of the animal's responses to the external environment, is mediated through the receptor organs, nervous system, and effector organs (muscles, glands, ciliated epithelia, etc.). Neurons, which are the basic cellular components of the nervous system, may be activated by nearly any stimulus that is intense enough. The specificity of nervous integration derives from the facts that the receptor cells of the body are attuned to specific environmental parameters (light, temperature changes, mechanical displacement, chemical change) and that they are activated by very slight changes in these parameters. The receptors, in turn, activate the neurons with which they are connected. These neurons have specific interconnections with other neurons and, eventually, with appropriate effectors.

Although distinct in many ways, hormonal and nervous integrations have much in common and tend to interact with each other. Both systems use chemical messengers to carry information either through the circulatory system or across the synapses between neurons. Parts of the brain, as we will see, influence the endocrine system, and the level of hormones in the blood affects many behavioral responses controlled by the nervous system.

Grossly, the nervous system comprises the **central nervous system**, which consists of the brain and spinal cord, and the **peripheral nervous system**, which is composed of the cranial, spinal, and autonomic nerves. Most of the nerves are mixed, being made up of long processes, called **axons**, of hundreds of sensory and motor neurons that carry impulses from the receptors to the brain and spinal cord and from the brain and cord to the effectors. Most of the interconnections of the sensory and motor neurons are made through **interneurons** located within the central nervous system, and this is where the integrative activities of the nervous system occur.

Some of the interconnections of sensory and motor neurons are quite direct, which results in a very rapid **reflex** response that is always the same under similar circumstances. For example, whenever you touch a hot stove, you jerk your hand away. However, other circumstances can produce different effects: sensory information may be stored, or it may be integrated with information coming in from other sense organs or it may be compared with information derived from past experience. Thus, a more complex response based on a totality of information, both present and past, might be generated in response to a particular stimulus.

A. REMOVING THE BRAIN

The brain can be studied by dissecting the brain of your specimen or a larger, preserved brain of a sheep. If you are dissecting the rat brain, first remove the skin from the top of the head between the eyes and external ears (auricles) as shown in Fig. 6-1. Cut ventrally through the skin in front of and behind each auricle so that you can push the auricles aside without detaching their bases. Cut away muscle tissue from the area between each eye and ear and from the back of the skull and the cranial part of the vertebral column. By using a pair of strong forceps, squeeze and break a part of the bony crest that forms the caudodorsal border of the skull. When a small hole has been made, enlarge it by carefully breaking away small pieces of bone from its edges. Do not push down on the brain any harder than necessary and leave the tough membrane (dura mater) covering the brain in place. Gradually uncover the top and sides of the brain and the cranial end of the spinal cord.

You should identify certain parts of the brain through the transparent dura mater as you proceed (Fig. 6-1). The **cerebellum** is located caudally and dorsally, and will be the first part exposed. It bears a pair of delicate lateral lobes called the **paraflocculi**, which you should try to save. The **medulla oblongata** lies caudal to the cerebellum and is continuous caudally with the spinal cord. The large, paired **cerebral hemispheres** lie rostral to the cerebellum; small, paired **olfactory bulbs** lie between the eyes, rostral to the cerebral hemispheres.

Avoid injuring the orbit and the skull at the base of the ear on one side, but continue to break away bone on the other side to work your way partly underneath the brain. As you do so, you will see some of the cranial nerves emerging from the ventrolateral surface of the brain. One group lies ventral to the paraflocculus. Cut the nerves, but leave stumps that are as long as possible attached to the brain. Carefully pick away the buttress of bone that extends between the cerebellum and cerebral hemispheres anterior to the paraflocculus. This is the **otic capsule,** and is contains the inner ear. Next cut through the caudal end of the medulla oblongata and gradually lift up the brain from its caudal end and from the side you have been undermining. Cut through the remaining cranial nerves and blood vessels, leaving as much tissue attached to the brain as possible. A dark-colored gland, the **hypophysis,** is attached to the middle of the ventral surface; try to leave it attached to the brain.

B. MENINGES

The brain is covered by three protective connective tissue layers known as the **meninges**. The tough, outermost layer is known as the **dura mater.** If you are studying a

Meninges

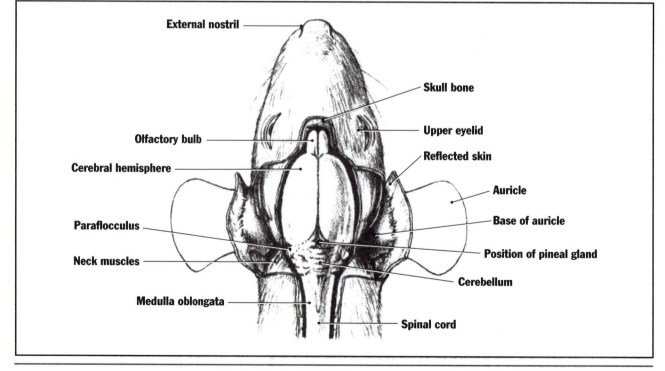

External nostril

Skull bone

Olfactory bulb

Upper eyelid

Cerebral hemisphere

Reflected skin

Auricle

Base of auricle

Paraflocculus

Position of pineal gland

Neck muscles

Cerebellum

Medulla oblongata

Spinal cord

FIGURE 6-1
Dorsal view of the head of a rat, showing the exposed brain
after reflecting the skin and removing some skull bones.

preserved sheep brain, it probably has been removed; you should look for it on a demonstration. If you are dissecting the rat, you have already identified the dura mater. Part of the dura mater will have adhered to the floor of the skull, but carefully peel off the rest of it without removing the stumps of the cranial nerves. A large venous sinus runs through the dorsal part of the dura mater and will be taken off with it, but do not remove the pineal gland, which lies dorsally between the cerebellum and cerebrum (the part of the brain consisting of the cerebral hemispheres) (Figs. 6-2B and 6-3A). A transverse septum of the dura mater, the **tentorium,** dips down between the cerebrum and cerebellum and helps to hold the brain in place; a smaller septum passes between the olfactory bulbs.

The other two meninges are difficult to separate grossly. A vascular **pia mater** closely invests the surface of the brain and dips into all its grooves and depressions. A nonvascular **arachnoid** covers the pia mater and does not dip into most of the depressions. In life, there is a **subarachnoid space** between the two meninges, which contains a lymphlike **cerebrospinal fluid** that circulates around the central nervous system and within the cavities of the brain and spinal cord. Cerebrospinal fluid is secreted by the vascularized pia mater into the cavities of the brain (see Section 1), and it drains back into the venous sinuses within the dura mater. Cerebrospinal fluid forms a liquid hydrostatic cushion around the brain and spinal cord, insulating these delicate structures from the surrounding bone of the skull and vertebrae and

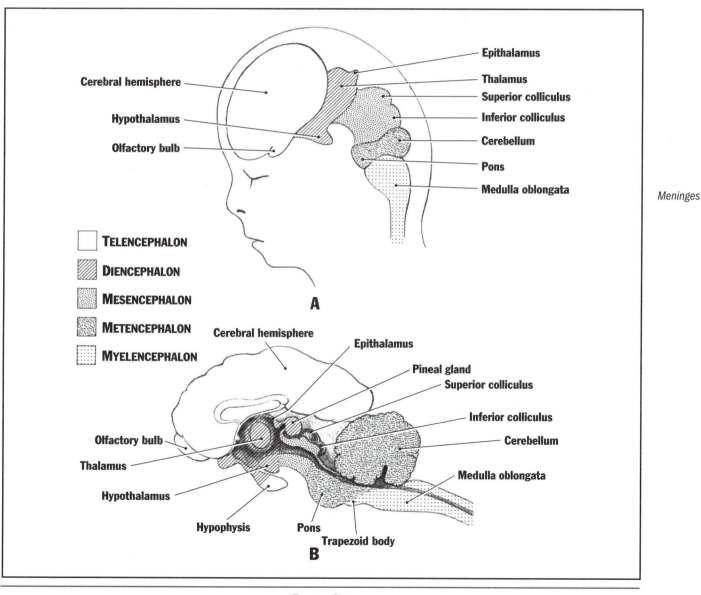

Meninges

TELENCEPHALON

DIENCEPHALON

MESENCEPHALON

METENCEPHALON

MYELENCEPHALON

FIGURE 6-2
Brain region. (A) A lateral view of an 8-week-old human embryo.
(B) A sagittal section of the brain of an adult sheep.

protecting them from mechanical injury. It also makes an important contribution to brain nutrition and drainage.

Arteries that supply the brain can be seen on its ventral surface. They receive blood primarily from the vertebral and internal carotid arteries (Exercise 4). Veins that drain the brain are tributaries of the internal and external jugular veins.

C. ENDOCRINE GLANDS ATTACHED TO THE BRAIN

Two major endocrine glands are associated with the brain. If you are dissecting the rat brain, you will have seen the **pineal gland** protruding between the two cerebral hemispheres and the cerebellum. The pineal gland of the sheep lies deep to the caudal end of the cerebral hemispheres (Fig. 6-2B). It can be seen by spreading the caudal ends of the cerebral hemispheres apart. The pineal gland develops embryonically as an outgrowth from a brain region known as the diencephalon. Because the diencephalon is located deep to the caudal parts of the cerebral hemispheres, you will not be able to see the narrow stalk by which the pineal attaches until later.

The pineal gland of mammals evolved from a light-receptive third eye found on the top of the head of some reptiles and earlier vertebrates. Indeed, its secretory cells are modified photoreceptors. These cells synthesize and release the hormone **melatonin,** particularly under dark conditions. Melatonin synthesis is inhibited by light. The seasonal enlargement of the ovary and testis in breeding rodents in the spring appears to result from the reduction of melatonin production as the length of the days increases. The pineal gland may also play a role in other seasonal and diurnal cycles. It has been proposed that jet lag occurs when the diurnal cycle of melatonin synthesis gets out of phase with the normal light–dark cycle.

The **hypophysis,** or pituitary gland, is a dark, ovular gland attached to the ventral surface of the diencephalon by a narrow stalk known as the **infundibulum** (Fig. 6-3B). The gland may have been left in the skull when the brain was removed; if it was not, you should detach it from the brain now. Do not destroy the pair of small nerves dorsal to it. One part of the hypophysis develops from the floor of the diencephalon and the other part develops as an ingrowth from the roof of the embryo's mouth. It produces, or stores, many hormones that have wide-ranging effects. **Melanophore stimulating hormone** promotes the synthesis and dispersion of the pigment melanin. **Thy-**

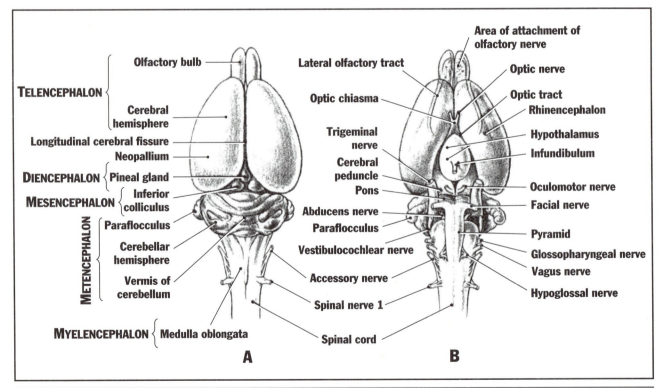

FIGURE 6-3
The brain of a rat. (A) Dorsal view. (B) Ventral view.

rotropic hormone stimulates the thyroid gland. **Adrenocorticotropic hormone** increases the activity of the adrenal cortex (Exercise 4). **Growth hormone** promotes body growth. Two **gonadotropic hormones** regulate the activity of the ovary and testis (Exercise 5). **Prolactin** stimulates the synthesis of milk by the mammary glands. **Oxytocin** helps release milk during lactation. **Antidiuretic hormone** promotes the reabsorption of water from parts of the kidney tubules. The last two hormones are actually synthesized in a part of the brain known as the hypothalamus, but they are stored and released by the hypophysis (see later).

D. BRAIN REGIONS

The various parts of the adult brain develop from five distinct enlargements in the early embryonic brain (Fig. 6-2): telencephalon, diencephalon, mesencephalon, metencephalon, and myelencephalon. It is convenient to apply the same terms to the adult brain. The **telencephalon** includes the **olfactory bulbs** and **cerebral hemispheres,** which you have seen. The **diencephalon** consists of the epithalamus, thalamus, and hypothalamus. The **hypothalamus** can be seen in a ventral view, lying caudal to the optic chiasma (Figs. 6-3B and 6-4). The pineal gland attaches to the epithalamus; you will see other parts of the epithalamus and the thalamus later (Section F), after you have cut off part of the cerebrum. The dorsal part of the **mesencephalon** bears two pairs of oval swellings: the rostral **superior colliculi** and caudal **inferior colliculi.** You can see the latter lying between the pineal gland and the cerebellum (Fig. 6-3A). The superior colliculi can be seen in sagittal sections (Figs. 6-2B and 6-6). Stalklike fiber tracts, known as the **cerebral peduncles,** are also a part of the mesencephalon, and they can be seen in a ventral view, lying caudal and lateral to the mamillary body. The **metencephalon** includes the **cerebellum** and, ventrally, the **pons,** which appears as a bundle of transverse fibers. The **myelencephalon** consists of the **medulla oblongata.**

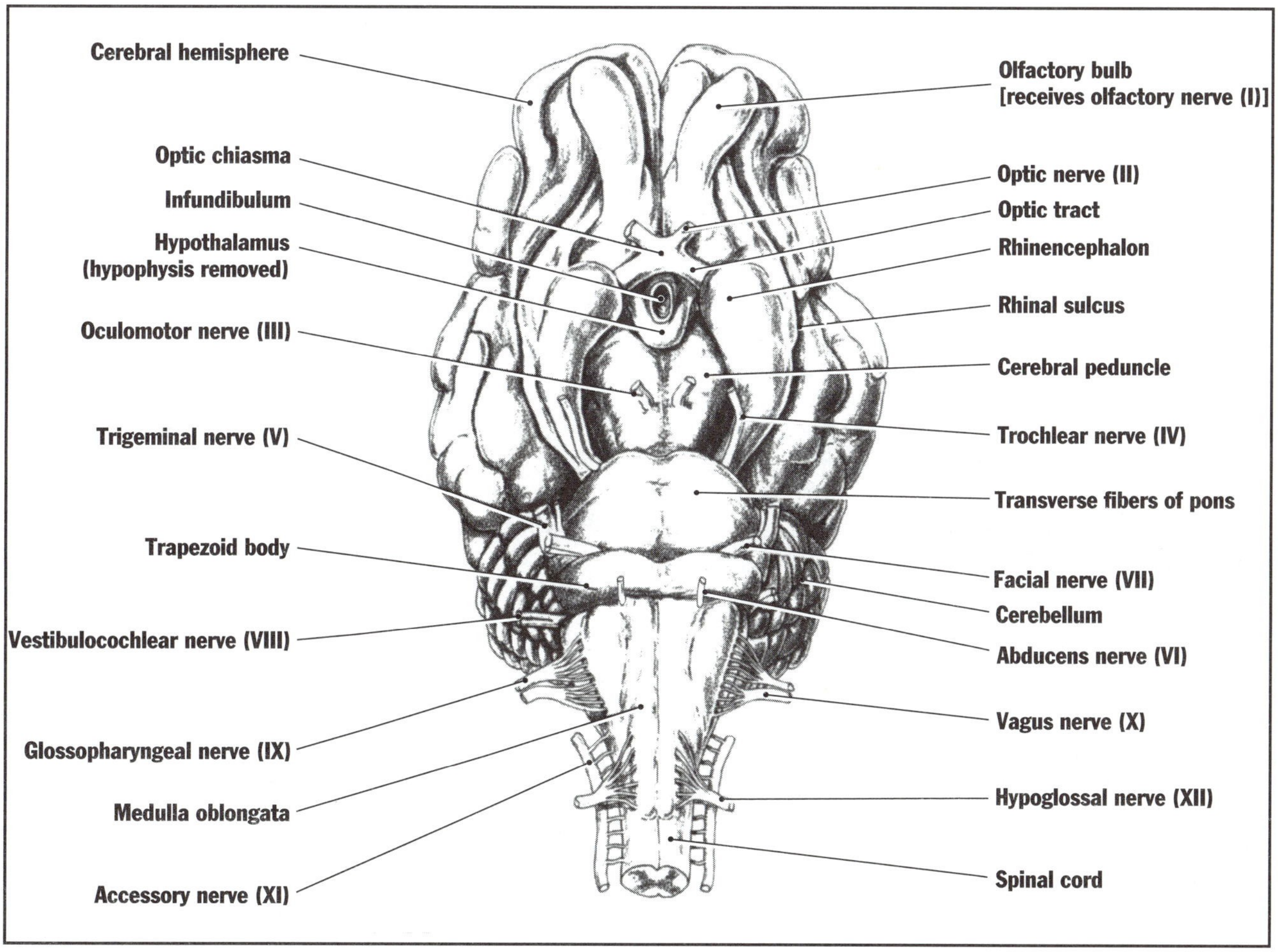
Cerebral hemisphere
Optic chiasma
Infundibulum
Hypothalamus
(hypophysis removed)
Oculomotor nerve (III)
Trigeminal nerve (V)
Trapezoid body
Vestibulocochlear nerve (VIII)
Glossopharyngeal nerve (IX)
Medulla oblongata
Accessory nerve (XI)
Olfactory bulb
[receives olfactory nerve (I)]
Optic nerve (II)
Optic tract
Rhinencephalon
Rhinal sulcus
Cerebral peduncle
Trochlear nerve (IV)
Transverse fibers of pons
Facial nerve (VII)
Cerebellum
Abducens nerve (VI)
Vagus nerve (X)
Hypoglossal nerve (XII)
Spinal cord

FIGURE 6-4
Ventral view of the brain of a sheep, with stumps of the cranial nerves.

E. CRANIAL NERVES

The stumps of the cranial nerves should be identified now, because some of them will be destroyed in the subsequent study of the brain. Most can be seen by examining the ventral and lateral surfaces of the rat or sheep brain (Figs. 6-3B and 6-4). The nerves are paired, and each has not only a name but also a number that corresponds to its relative position in the brain of a human being.

I. Olfactory Nerve. Consists of many sensory fibers coming from the olfactory cells in the nasal cavity to enter the ventral surface of the olfactory bulb. These fibers are usually torn off when the brain is removed from the skull.

II. Optic Nerve. Consists of sensory neurons coming from the retina. Because the retina develops as an outgrowth of the brain, the optic nerve is technically a brain fiber tract and not a typical peripheral nerve. One-half of the fibers in each nerve (those from the caudal or lateral part of the retina) terminate in the side of the brain they entered; the other half cross to the opposite side of the brain, and this crossing forms the **optic chiasma.** Thus images from each eye are projected to each side of the brain. This crossover of visual information plays a role in stereoscopic vision and results in depth perception. Most impulses from other sense organs also cross at some point in their ascent to higher brain centers, but not all of these decussations are as superficial as the optic chiasma, and, therefore, are not as easily seen.

III. Oculomotor Nerve. A small nerve attaching to the cerebral peduncle. It sends motor impulses to four of the seven small, straplike muscles that move the eyeball (extrinsic ocular muscles), to the ciliary muscles within the eyeball that focus the eye (see Exercise 7), and to a small muscle that raises the upper eyelid. Like most motor nerves, it also returns a few sensory, proprioceptive impulses that provide information on the degree and extent of muscle contraction.

IV. Trochlear Nerve. A small nerve that may be seen later on the roof of the mesencephalon. Its motor fibers supply one of the extrinsic ocular muscles.

V. Trigeminal Nerve. A large, mixed nerve (i.e., sensory and motor) attaching to the underside of the metencephalon lateral to the pons. It contains sensory fibers that carry impulses from the skin of the face and head and motor fibers that carry impulses to the jaw muscles. Jaw muscles are branchiomeric muscles, for they evolved from first visceral (branchial) arch muscles, which are associated with the mouth and pharynx of ancestral fishes (Exercise 2).

VI. Abducens Nerve. A small nerve attaching ventrally to the rostral part of the medulla oblongata. It contains motor fibers that go to two extrinsic ocular muscles.

VII. Facial Nerve. A mixed nerve lying just caudal to the base of the trigeminal nerve. It contains motor fibers that carry impulses to the facial muscles that move the skin on the head and face, to motor fibers that carry impulses to the tear glands, and to some of the salivary glands, and sensory fibers that carry impulses from the taste buds on the anterior two-thirds of the tongue. The muscles that are supplied by this nerve are branchiomeric, having evolved from those of the second visceral arch of ancestral fishes (Exercise 2).

VIII. Vestibulocochlear Nerve. Attaches to the lateral surface of the rostral part of the medulla oblongata. Its sensory fibers carry impulses related to equilibrium from the vestibular parts of the ear (e.g., the semicircular ducts) and auditory impulses from the cochlea of the ear.

IX. Glossopharyngeal Nerve. A mixed nerve that attaches to the medulla oblongata caudal to the vestibulocochlear nerve, by several fine rootlets. It consists of motor fibers that carry impulses to pharyngeal muscles and certain salivary glands and of sensory fibers that carry impulses from the caudal part of the tongue and the pharyngeal lining. These pharyngeal muscles are branchiomeric having evolved from muscles of the third visceral arch.

X. Vagus Nerve. A mixed nerve that attaches to the medulla oblongata by fine rootlets located caudal to those of the glossopharyngeal nerve. You have already seen the peripheral part of the vagus beside the common carotid artery and on the thoracic part of the esophagus (Exercise 4). Its motor fibers supply part of the pharynx, larynx, heart, stomach, and intestinal region; its sensory fibers return impulses from the larynx, lungs, heart, and stomach. The remaining pharyngeal and laryngeal muscles are also branchiomeric, having evolved from muscles of the caudal visceral arches.

The motor fibers in this nerve, which go to the heart and intestine, as well as the motor fibers in the third, seventh, and ninth nerves, which go to the ciliary muscles, tear glands, and salivary glands, respectively, belong to the **parasympathetic division** of the **autonomic nervous system.**

XI. Accessory Nerve. Attaches by a series of rootlets to the caudolateral part of the medulla oblongata and to the cranial part of the spinal cord. It carries motor impulses to certain neck and shoulder muscles (the sternocleidomastoid and trapezius groups) that evolved from caudal visceral arch muscles of ancestral fishes (Exercise 2).

Cranial Nerves

XII. Hypoglossal Nerve. Attaches by several rootlets to the caudoventral surface of the medulla oblongata. It carries motor impulses to the muscles of the tongue.

F. BRAIN STRUCTURE AND FUNCTIONS

We will begin our study of brain structure and functions by examining the **telencephalon,** the most rostral of the five brain regions we described earlier (Fig. 6-2). Recall that the telencephalon includes the olfactory bulbs and cerebral hemispheres. Olfactory cells from the nasal cavities terminate in the **olfactory bulbs.** From these, the impulses are carried by other neurons to the ventral part of the cerebral hemispheres, known as the **rhinencephalon.** One of the pathways, the **lateral olfactory tract,** appears as a whitish line on the underside of a cerebral hemisphere (Fig. 6-3B). The rhinencephalon is an olfactory integration center. It comprises nearly all the cerebral hemispheres in nonmammalian vertebrates, but, in the course of the evolution of mammals, other sensory and motor centers evolved in the cerebral hemispheres and caused them to expand and grow caudally over the diencephalon and mesencephalon. The new part of the cerebrum that has evolved in mammals is called the **neopallium,** and it forms the laterodorsal part of each cerebral hemisphere.

Rhinencephalon and neopallium are clearly separated in the sheep brain by a prominent, horizontal groove known as the **rhinal sulcus** (Fig. 6-4), but this groove is not present in the rat brain. The surface of the neopallium of the cerebral hemispheres of the rat and many other small mammals is smooth, but the surface of the brain of the sheep and many other large mammals is thrown into a complex pattern of ridges, the **gyri,** separated by grooves, the **sulci.** The cell bodies of most cerebral neurons lie in or near the surface, and the gyri and sulci increase the surface area needed to accommodate them.

Break the arachnoid membrane that connects the cerebral hemispheres dorsally and separate them enough to be able to see into the deep longitudinal groove between them, which is called the **longitudinal cerebral fissure.** At the base of this fissure, a wide, transverse band of fibers, the **corpus callosum,** can be seen. This commissure connects the two cerebral hemispheres and allows the exchange of information between them. Cut through the corpus callosum and then make a transverse incision through one cerebral hemisphere, caudal to the optic chiasma (Fig. 6-5), so that you can remove the part of the cerebral hemisphere that overlaps the diencephalon and mesencephalon.

Notice that the cut surface of a cerebral hemisphere is composed of a superficial **cortex** of **gray matter** and that most of the deeper parts of the cerebral hemisphere consist of **white matter.** The gray matter consists of the cell bodies and unmyelinated fibers of neurons; the white matter consists largely of the long, myelinated processes of neurons (axons) that form fiber tracts. The axons in the white matter belong to many groups of neurons that cannot be distinguished grossly, but you should know what they represent collectively (see Fig. 6-8). **Sensory fibers** project from the brain stem

Brain Structure and Functions

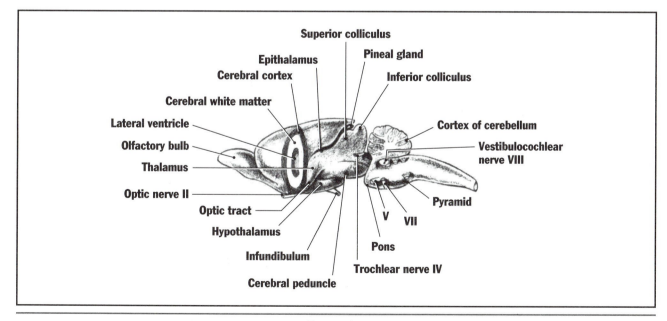

FIGURE 6-5
Lateral view of the brain of a rat after parts of the cerebrum and cerebellum have been removed.

and bring sensory information to specific cerebral regions. **Association fibers,** which interconnect different parts of the same cerebral hemisphere, and **commissural fibers,** which interconnect the two cerebral hemispheres, make it possible for different sorts of sensory information to be integrated and compared with memories of past experience. Eventually, motor impulses are initiated in specific motor regions of the cerebral cortex and sent out on **motor fibers** to the brain stem and to the spinal cord.

In the cut surface of the cerebrum, also notice one of the cavities of the brain, the **lateral ventricle.** Spread the ventricle open slightly, and look for a small tuft of tissue along its ventral border. This tissue, which is known as a **choroid plexus,** is a membrane composed of the vascular pia mater united with the ependymal epithelium that lines all the cavities within the brain and spinal cord. It secretes the cerebrospinal fluid into these cavities. You will see other choroid plexuses later.

Make a sagittal section of the brain, keeping as close to the median longitudinal plane as possible (Figs. 6-6 and 6-7). Several bundles of commissural fibers that interconnect the two sides of the telencephalon can be seen in a sagittal section. The largest and most dorsal of these commissures is the **corpus callosum,** which connects the nonolfactory parts of the cerebral hemispheres. The wall of the cerebral hemisphere ventral to this is the thin **septum pellucidum.** By breaking through the septum pellucidum, you will see another view of a lateral ventricle. There is one lateral ventricle in each cerebral hemisphere, and they represent the first and second ventricles.

Another band of white fibers, the **fornix,** lies ventral to the septum pellucidum. It curves rostrally and ventrally toward the hypothalamus, or floor of the diencephalon, but soon turns laterally and disappears from

the sagittal plane. The fornix is part of a neural pathway that interconnects olfactory centers in the telencephalon with the hypothalamus and thalamus. This neural pathway and associated parts of the brain form the **limbic system.** The limbic system influences many aspects of motivational and emotional behavior related to self- and species preservation, including feeding, drinking, fighting, and reproduction. This system also has been implicated in the formation of short-term memories.

Near the point where the rostral end of the fornix disappears from the sagittal plane is a small, transverse bundle of fibers with a round cross section, the **anterior commissure,** which connects the major olfactory parts of the two cerebral hemispheres. A thin vertical partition extends from the anterior commissure to the crossing of the optic nerves, the optic chiasma, on the ventral surface of the brain. This is the **lamina terminalis,** the most rostral part of the embryonic brain in the midline. The cerebral hemispheres develop as anterolateral expansions from this region. As noted earlier, a partial decussation of optic fibers occurs in the optic chiasma.

The cerebral hemispheres of nonmammalian vertebrates are relatively small and are situated entirely rostral to the other regions of the brain. As they assume additional functions in the course of evolution, they enlarge and grow caudally over the dorsal and lateral surfaces of much of the rest of the brain. The overgrowth that has occurred is evident in the sagittal section. The major brain region caudal to the telencephalon, which is largely covered by the cerebral hemispheres, is the **diencephalon.** It extends from the fornix, anterior commissure, and lamina terminalis caudally to include the pineal gland. The diencephalon contains a narrow chamber, the **third ventricle,** whose extent can be recognized by its shiny lining of ependymal epithe-

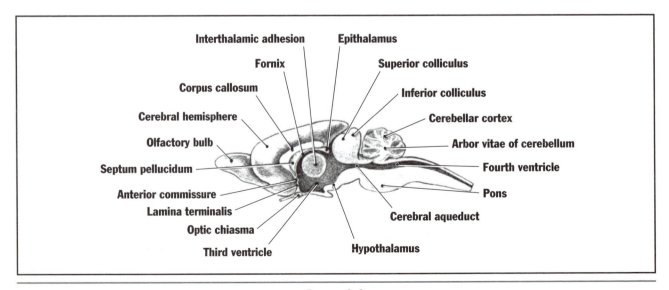

FIGURE 6-6
Sagittal section of the brain of a rat.

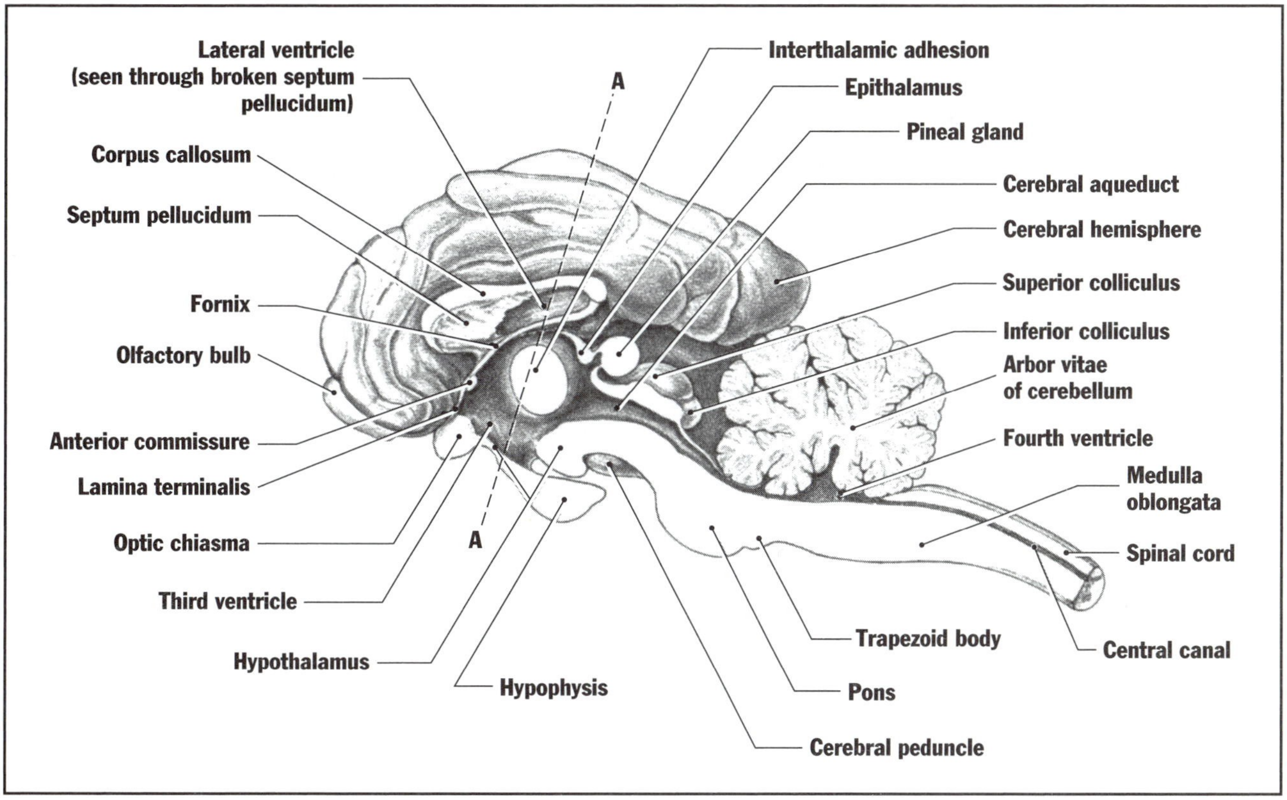
Lateral ventricle (seen through broken septum pellucidum)
Corpus callosum
Septum pellucidum
Fornix
Olfactory bulb
Anterior commissure
Lamina terminalis
Optic chiasma
Third ventricle
Hypothalamus
Hypophysis
A
A
Interthalamic adhesion
Epithalamus
Pineal gland
Cerebral aqueduct
Cerebral hemisphere
Superior colliculus
Inferior colliculus
Arbor vitae of cerebellum
Fourth ventricle
Medulla oblongata
Spinal cord
Central canal
Trapezoid body
Pons
Cerebral peduncle

FIGURE 6-7
Sagittal section of the brain of a sheep.

Brain Structure and Functions

lium (Figs. 6-6 and 6-7). (Further dissection may be necessary to expose the third ventricle.) This ventricle is crossed by a large circular mass of nervous tissue, the **interthalamic adhesion.** Inconspicuous **interventricular foramina** connect the two lateral ventricles with the third ventricle.

The diencephalon can be divided into the **epithalamus,** lying dorsal to the third ventricle; the **hypothalamus,** ventral to it; and the halves of the **thalamus** on each lateral side of the third ventricle. The epithalamus consists of the vascular roof of the third ventricle, the pineal gland, and adjacent nervous tissue related to olfaction. Much of the roof of the diencephalon is a choroid plexus, and it is often destroyed in making a sagittal section.

The hypothalamus is the conspicuous oval area that can be seen on the ventral surface of the brain posterior to the optic chiasma (Figs. 6-3 and 6-4). The hypophysis, or pituitary gland, attaches to this area by a stalk called the infundibulum. One part of the hypophysis develops embryonically as an outgrowth from the diencephalon, and the other part as an outgrowth from the roof of the mouth. The hypothalamus is concerned with the integration of many autonomic functions: sleep, body temperature, water balance, appetite, and carbohydrate and fat metabolism. It exerts its influence by means of motor pathways that project away from it and by producing various hormones. As we have mentioned, two hypophyseal hormones (oxytocin and antidiuretic hormone) are synthesized by specialized hypothalamic neurons and travel along their axons through the infundibulum to be stored in and eventually released from the hypophysis. The synthesis and release of other hypophyseal hormones is regulated by a series of **releasing** and **inhibiting hormones** that are produced in the hypothalamus. These hypothalamic hormones then travel along the infundibulum in a series of minute blood vessels (the **hypophyseal portal system**) to their target cells.

The thalamus is very large in mammals. It includes the interthalamic adhesion and the entire lateral wall of the diencephalon. Its lateral surface can be seen on the brain half from which you removed the caudal part of a cerebral hemisphere (Fig. 6-5). Strip off the remaining meninges from the lateral surface of the thalamus and the adjacent mesencephalon, looking for the trochlear nerve as you do so. Also notice the **optic tract,** which consists of optic fibers extending from the optic chiasma to the thalamus. The thalamus, like the cerebrum, has increased in importance with the evolution from ancestral vertebrates to mammals. All sensory information, apart from olfaction, entering the spinal cord or brain on sensory neurons proceeds to the thalamus on interneu-

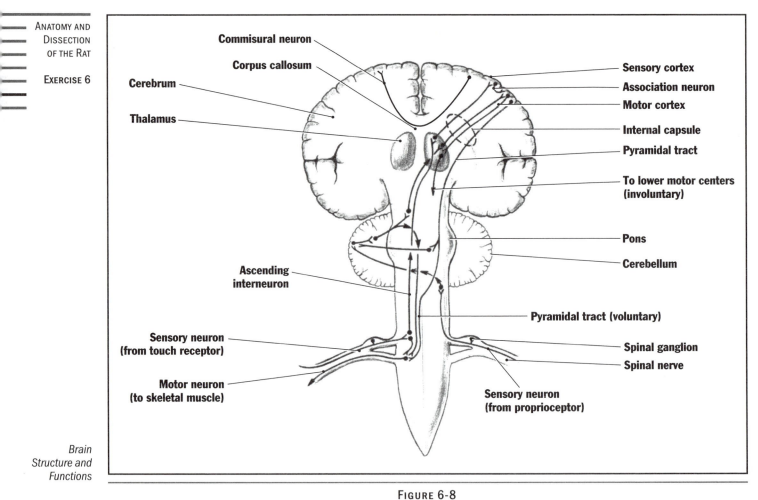

Commisural neuron

Corpus callosum

Cerebrum

Thalamus

Sensory cortex

Association neuron

Motor cortex

Internal capsule

Pyramidal tract

**To lower motor centers
(involuntary)**

Pons

Cerebellum

**Ascending
interneuron**

Pyramidal tract (voluntary)

**Sensory neuron
(from touch receptor)**

Spinal ganglion

Spinal nerve

**Motor neuron
(to skeletal muscle)**

**Sensory neuron
(from proprioceptor)**

FIGURE 6-8

Ventral view of a wiring diagram of the nervous system, showing interconnections of neurons in some
of the major sensory and motor pathways. Each dark, round spot represents the cell body of a neuron,
and lines extending from them represent axons. Nerve impulses travel away from the cell bodies.

rons (Fig. 6-8). On the way, most of these decussate, or cross to the opposite side of the body. Sensory information is sorted in the thalamus, and some or all of it is relayed on other interneurons that contribute to a fiber tract known as the **internal capsule** as they pass through the thalamus and fan out through the base of the cerebrum to specific sensory areas of the cerebral cortex. Some involuntary motor activity initiated in the cortex is also relayed in the thalamus on its way to lower motor centers. But the thalamus is more than a switchboard. Although many of its functions are as yet undiscovered, there is much evidence that the thalamus interacts with the cerebral cortex in many of the higher mental processes.

The third major region of the brain, the **mesencephalon,** lies between the diencephalon and the large cerebellum. Examine the mesencephalon in a lateral view (Fig. 6-5). The **cerebral peduncles** on its floor are part of the pyramidal tract, a motor pathway (Fig. 6-8) that extends directly from motor centers in the cerebrum

through the internal capsule to the origins of the cranial and spinal nerves without relaying in the thalamus. Impulses for many voluntary movements pass through it. Like ascending sensory pathways, most of the descending pyramidal fibers decussate before reaching the motor neurons of cranial and spinal nerves.

Dorsally, the mesencephalon bears two pairs of roundish swellings. The larger and more rostral pair are the two **superior colliculi;** the smaller and more caudal pair, the two **inferior colliculi** (Figs. 6-6 and 6-7). The superior colliculi are also known as the optic lobes, especially in nonmammalian vertebrates. The optic lobes are the major integration centers in the brains of fishes and amphibians, but in mammals most of the integrative functions of the optic lobes have been transferred to the cerebrum. However, a few optic fibers extend beyond the thalamus to end in the superior colliculi. The superior colliculi retain their function as centers for certain eye reflexes, including the pupillary reflex, accommodation, and eyeball movements. Certain auditory reflexes occur in the inferior colliculi. You may be able to see in the sagittal section that a small cerebral aqueduct runs through the mesencephalon and connects the third ventricle to the fourth ventricle, which starts in the metencephalon.

The large **cerebellum** and the ventral part of the brain to which it attaches constitute the fourth major brain region, the **metencephalon** (Figs. 6-6 and 6-7). The cerebellum resembles the cerebrum of the sheep in being composed of many folds, the **folia,** and in having a gray cortex. Its white matter forms a treelike pattern called the **arbor vitae.** The cerebellum can be divided into several parts, the most conspicuous of which are a median **vermis** and paired lateral **cerebellar hemispheres** (Fig. 6-3A). The paraflocculi are part of the cerebellar hemispheres.

The primary sensory inflow of the cerebellum is from the part of the inner ear concerned with equilibrium and from the proprioceptive organs in muscle that record the degree and state of muscular contraction. It also receives copies, so to speak, of motor directives from the cerebellum (Fig. 6-8). These terminate in the ventral part of the metencephalon, the **pons.** Neurons receiving these impulses decussate in the **transverse fibers of the pons** and lead to the cerebellar cortex. The cerebellum is therefore a crucial center for muscular coordination, monitoring the orientation and motor activity of the body, and initiating corrective impulses that either go back to the cerebrum by way of the thalamus or go directly to the cell bodies of the motor neurons. Several cranial nerves attach just caudal to the pons, and their stumps can be found on good specimens (see Figs. 6-3B and 6-4): the trigeminal, abducens, facial, and vestibulocochlear nerves.

The fifth and final region of the brain is the **myelencephalon** (Figs. 6-2, 6-3, and 6-4). It consists of the **medulla oblongata,** which merges caudally with the spinal cord. At the front of the underside of the medulla oblongata, just caudal to the pons, there is a small transverse band of fibers, the **trapezoid body.** This is an acoustic commissure. It is more evident in the sheep than in the rat brain. The pyramidal motor system forms a pair of bulges, the **pyramids,** on the ventral surface of the medulla oblongata caudal to the trapezoid body. The fourth ventricle continues from the metencephalon into the medulla oblongata and has a very thin, vascular roof that forms another choroid plexus. Microscopic perforations in the roof of the fourth ventricle permit cerebrospinal fluid to escape from this ventricle and to circulate between the pia mater and arachnoid membrane. Many visceral activities are regulated by centers in the medulla oblongata: rate of heart beat, blood pressure, breathing movements, salivation, and swallowing. Stumps of the remaining cranial nerves can be found attached to the surface of the medulla oblongata on good specimens: the glossopharyngeal, vagus, accessory, and hypoglossal nerves.

G. SPINAL CORD, SPINAL NERVES, AND AUTONOMIC NERVOUS SYSTEM

You saw the cranial end of the spinal cord and perhaps the first spinal nerve if you removed the rat brain from the skull (Figs. 6-1 and 6-3). The structures of the spinal cord and nerves can be seen most clearly by examining microscope slides of these organs from a frog or mammal (Fig. 6-9). The ventral surface of the **spinal cord** can be recognized by a conspicuous **ventral fissure.** As in all chordates, the spinal cord of vertebrates is hollow and contains a small **central canal,** which, like the ventricles of the brain, is lined by a nonnervous **ependymal epithelium.** A small amount of lymph-like cerebrospinal fluid circulates in the central canal. The rest of the spinal cord and the spinal nerves are composed primarily of elongated neurons. They are distributed in such a way that one can recognize a centrally located, somewhat butterfly-shaped **gray matter** and a peripheral **white matter.** These terms derive from the color of these areas in fresh preparations; with certain stains, the gray matter may appear lighter than the white matter on the slides. Central gray matter and peripheral white matter continue forward through much of the brain, although, as you have already seen, some gray matter migrates superficially to form the gray cortex of the cerebellum and cerebrum.

The gray matter contains the cell bodies of neurons and fibrous processes that are not enveloped by a fatty myelin sheath consisting mostly of lipids. The **cell bodies** of the **motor neurons** form a conspicuous group in the ventrolateral part of the gray matter (Fig. 6-9A). They are large, triangular or diamond-shaped cells with a conspicuous nucleus containing a prominent, eccentric nucleolus. Nerve impulses from other neurons are received directly by these cell bodies or by short processes known as **dendrites** that attach to most of the angles of the cell bodies. Impulses then travel along long processes, the **axons,** one of which leaves each cell body, passes through the **ventral root** of a **spinal nerve,** and is distributed to the skeletal muscles of the body (Fig. 6-9B).

The white matter consists of myelinated axons of neurons that carry impulses from the spinal cord to centers in the brain or from the brain down to the motor neurons. These are seen in cross sections. The few small nuclei that you may see among them belong to cells that ensheath the axons and deposit the myelin to connective tissue, or to cells in the walls of blood vessels.

Sensory neurons from receptor cells enter the **dorsal root** of a **spinal nerve;** their cell bodies are located in the **spinal ganglion** (Fig. 6-9A). From there, impulses travel along sensory axons into the spinal cord. Some ascend in the white matter to the brain. Others enter the dorsal part of the gray matter where they may terminate and be relayed elsewhere by interneurons, whose cell bodies are difficult to distinguish, or they may continue through the gray matter to terminate on the

Spinal Cord,
Spinal Nerves,
and
Autonomic
Nervous
System

Spinal Cord,
Spinal Nerves,
and
Autonomic
Nervous
System

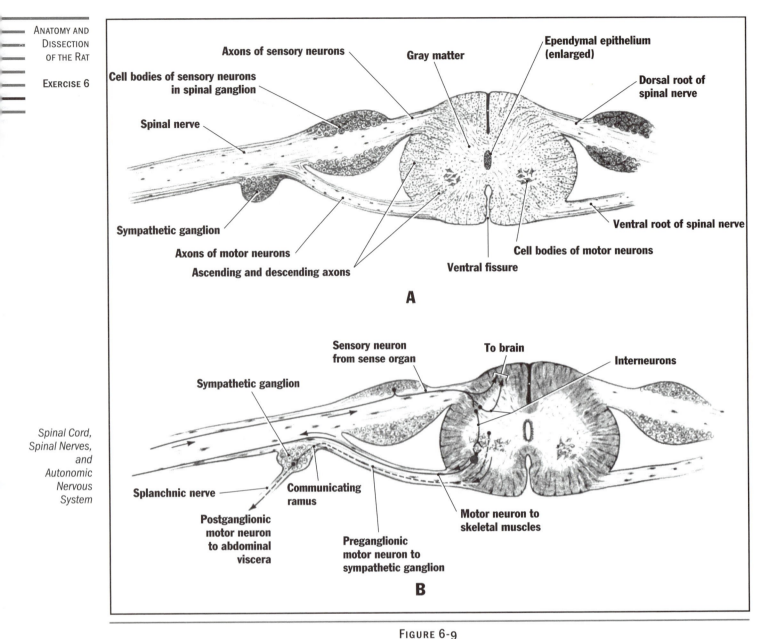

Cell bodies of sensory neurons in spinal ganglion
Axons of sensory neurons
Gray matter
Ependymal epithelium (enlarged)
Dorsal root of spinal nerve
Spinal nerve
Sympathetic ganglion
Axons of motor neurons
Ascending and descending axons
Ventral fissure
Cell bodies of motor neurons
Ventral root of spinal nerve

A

Sympathetic ganglion
Sensory neuron from sense organ
To brain
Interneurons
Splanchnic nerve
Communicating ramus
Postganglionic motor neuron to abdominal viscera
Preganglionic motor neuron to sympathetic ganglion
Motor neuron to skeletal muscles

B

FIGURE 6-9
Spinal cord and spinal nerve of a frog. (A) Drawing of a transverse section.
(B) Diagram of the major types of neurons and their interconnections.

cell bodies of the motor neurons. The termination of a sensory neuron either directly on a motor neuron or indirectly through an interneuron is responsible for simple **spinal reflexes** (Fig. 6-9B).

In some slides, a **sympathetic ganglion** may be seen attached to the ventral surface of a spinal nerve a short distance to the union of the roots of the spinal nerve. Sympathetic ganglia lie on the sympathetic cord. The sympathetic cord and its cervical extension into the neck are very small in the rat, but you may have seen them during the dissection of blood vessels beside which they lie (Exercise 4). The neurons in the sympathetic cord and ganglia belong to the **sympathetic division** of the

autonomic nervous system. You have previously seen the vagus nerve (Section E of this Exercise, Exercises 3 and 4), which contains neurons belonging to the parasympathetic division of the autonomic nervous system. A unique feature of autonomic innervation is that the first motor neuron, whose cell body lies in the gray matter of the brain or spinal cord, always ends in some peripheral **ganglion,** where it relays with a second motor neuron that continues to the organ being supplied. Sympathetic ganglia are located on the sympathetic trunk and at the base of the coeliac and cranial mesenteric arteries (Exercise 4); parasympathetic ganglia are usually located within the wall of the organ supplied (Exercise 3).

Parasympathetic and sympathetic stimulation have mutually antagonistic effects on most of the organs they supply. In general, sympathetic stimulation provides more energy to the skeletal muscles of the body by increasing the rate and force of heart contraction, increasing blood pressure, sending more blood to active skeletal muscles (and less to gut muscles), and increasing the blood sugar level. This complex of activities is often called the **"flight or fight"** reaction. Parasympathetic stimulation has the opposite effects, promoting the conservation and restoration of energy. It is a **"rest and digest"** reaction. Heart rate and force of heart contractions decrease, blood pressure falls, more blood is directed to the digestive organs, sugars are stored.

In addition to its sympathetic and parasympathetic divisions, the autonomic nervous system also includes an **intramural nervous system,** which is confined to the gut wall. It consists mostly, but not entirely, of postganglionic neurons of the parasympathetic division of the autonomic nervous system. The cell bodies of these neurons are found in clusters called **intramural ganglia** (see Exercise 3). The intramural nervous system integrates peristaltic and other activities of the gut. It is autonomous to a large extent, but is modulated by sympathetic and parasympathetic stimulation.

Spinal Cord, Spinal Nerves, and Autonomic Nervous System

Nervous Coordination: Sense Organs

R ECEPTOR CELLS MAY BE MODIFIED NEURONS, such as the olfactory cells of the nose, or specialized cells, such as the rods and cones of the retina in the eye. Upon activation, receptors initiate nerve impulses in sensory neurons. There are many types of receptors specialized to respond to minute changes in particular environmental parameters. Chemoreceptors in the nose and on the tongue respond to changes in the chemical environment; mechanoreceptors in the skin, muscles, and ear respond to slight mechanical deformations in tissues; photoreceptors in the eye respond to changes in the intensity and wavelength characteristics of light; and thermoreceptors respond to changes in temperature and infrared radiation. Receptor cells by themselves cannot be seen without using high magnification, but they may be aggregated along with other tissues to form grossly visible sense organs such as the eye, ear, and nose. The associated tissues support and protect the receptor cells and may amplify the stimulus and direct it toward the receptors. We will examine the eye, ear, and nose of the rat as examples of sense organs. Study them on the side of the head that was least affected by removing the brain.

A. EYE

The eye is a complex sense organ attuned to light and to changes in the visual field. The sense organ itself, the **eyeball,** is lodged within a socket of the skull, the **orbit,** and is surrounded by various accessory structures that protect, moisten, and cleanse it. You have already seen the movable upper and lower **eyelids,** and the **nictitating membrane** at the rostral edge of the eye (Exercise 1). Notice them again. The nictitating membrane can slide across the eyeball, which helps to keep its surface clean. The surface of the eye is kept moist by the secretions of a group of tear, or lacrimal, glands; one of these glands, the **extraorbital lacrimal gland,** you observed when you dissected the salivary glands (Exercise 3). Find it again beneath the skin, ventral to the base of the ear (Fig. 7-1), and follow its duct toward the caudodorsal corner of the orbit. Dissect away the lower eyelid and cut through the delicate membrane, the **conjunctiva,** that reflects from the inside of the eyelid to cover the surface of the eyeball. The triangular-shaped **intraorbital lacrimal gland** can now be seen lying caudal and ventral to the eyeball. Its inconspicuous duct joins that of the extraorbital gland, and together they open near the caudal union of the upper and lower eyelids. Dissect between the intraorbital lacrimal gland and the eyeball to find the deeper **Harderian gland** (accessory lacrimal gland). It extends forward, ventral to the eyeball, curves dorsally deep to the nictitat-

Eye

ing membrane, and then turns caudally over the dorsal surface of the eyeball. Its inconspicuous duct opens underneath the nictitating membrane, and its secretions facilitate movements of this membrane. A **nasolacrimal duct,** which you will not find, carries tears from the anterior corner of the eyelids into the nasal cavity.

Cut away the upper eyelid. As you do so, notice the small straplike muscle, the **levator palpebrae superioris,** that attaches to the eyelid and raises it. This muscle is one of the **extrinsic ocular muscles.** Pick away the Harderian gland and you will find the other extrinsic ocular muscles, which attach to the eyeball and move it. They are too small to be identified individually. Pick them away and find the **optic nerve** leaving the medial or back side of the eyeball.

Dissect the eyeball of your rat specimen or a larger eyeball from a sheep or cow. If you study a larger eyeball, first remove the fat and extrinsic ocular muscles and find the optic nerve, which attaches to the medial side of the eyeball slightly rostral and ventral to its center. This will help you orient the specimen in space. The eyeball is composed of three layers (Fig. 7-2). The outermost, a dense fibrous layer, is the **fibrous tunic.** The posterolateral (deep) part of the fibrous tunic is an opaque **sclera;** but, at the front of the eyeball, the fibrous tunic is a transparent **cornea.** Although the cornea is rather cloudy in preserved specimens, you can usually look through it and see the pigmented **iris,** with a circular opening, called the **pupil,** in its center.

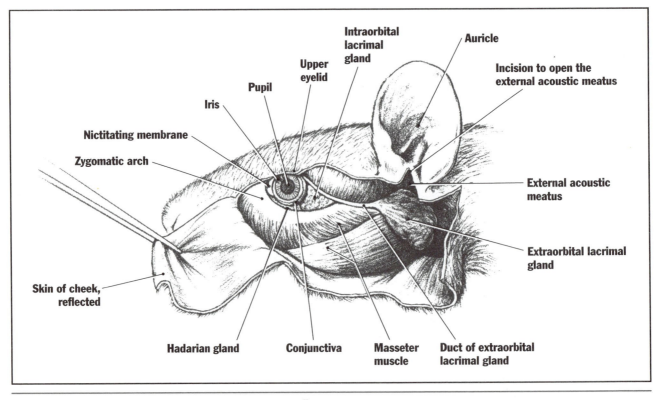

FIGURE 7-1
Lateral view of the eye and ear of a rat.

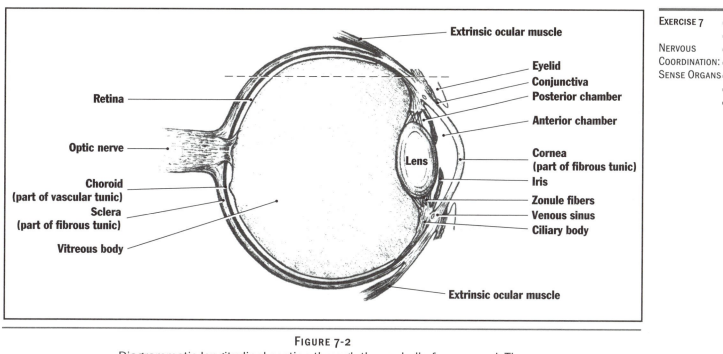

Labels on figure:
- Extrinsic ocular muscle
- Eyelid
- Conjunctiva
- Posterior chamber
- Anterior chamber
- Retina
- Optic nerve
- Cornea (part of fibrous tunic)
- Iris
- Lens
- Choroid (part of vascular tunic)
- Sclera (part of fibrous tunic)
- Vitreous body
- Zonule fibers
- Venous sinus
- Ciliary body
- Extrinsic ocular muscle

FIGURE 7-2
Diagrammatic longitudinal section through the eyeball of a mammal. The
plane for removing a tangential section is shown by the dashed line.

With a pair of fine scissors, cut off approximately the dorsal one-quarter of the eyeball, including the top of the cornea and iris. Continue cutting through the sclera. Before dissecting further, look into the eyeball. Note again the outer fibrous tunic (Fig. 7-2). A dark **choroid,** a part of the **vascular tunic,** lies internal to the fibrous tunic, and the whitish **retina** lies internal to the choroid. The image falls on this innermost layer. The retina is not attached to the choroid but is held against it by the gelatinous **vitreous body,** which fills the large space between the **lens** near the front of the eyeball and the back wall of the eyeball.

Before dissecting further, submerge the eyeball in a dish of water so that delicate structures will be supported. Now make a vertical, equatorial cut around the center of the eyeball. This will divide the eyeball into front, or anterior, and back, or posterior, halves (Fig. 7-3). Carefully remove the vitreous body, and you will see the retina in the posterior half of the eyeball more clearly (Fig. 7-3A). It may have floated away from the choroid. The whitish or grayish layer you are looking at is the **nervous layer of the retina.** Embryonically, there is also a **pigmented layer of the retina,** but this becomes attached to the choroid in the adult. Because the retina develops as an outgrowth of the embryonic brain, its nervous layer contains several layers of neurons with complex interconnections between them.

The photoreceptive cells, the **rods** and **cones,** lie in the retina next to the choroid, consequently light entering the eye must pass first through the neuronal layers to reach and stimulate them. Rods are activated by weak light, thereby allowing vision in dimly lighted areas; cones are activated by brighter light. Because there are three kinds of cones, with respective light absorption maxima in the red, green, and blue parts of the light spectrum, cones also mediate color vision.

Considerable processing of the image occurs within the retina before neuronal signals continue along the optic nerve to the brain. The retina is firmly attached to the wall of the eyeball only at a small circular area, the **optic disk,** where its neurons leave the retina to form the optic nerve. Rods and cones are absent in this area, so the optic disk is a "blind spot," and an image that falls on the optic disk cannot be perceived.

Let the retina float away from the wall of the eyeball. Most of the vascular tunic is a black choroid, which you have been looking at. It lies between the sclera and the retina. The choroid is rich in blood vessels, which help nourish the retina. Its pigment, which is derived from the pigmented layer of the embryonic retina, absorbs scattered light (like the black color inside a camera) and, thus, prevents a blurring of the image falling on the retina. In the eyes of some mammals, such as the sheep and cow, part of the choroid is modified and has an iridescent sheen. This area is the **tapetum lucidum.** The tapetum lucidum reflects some of the light passing through the retina back onto the photoreceptive cells, thereby facilitating vision under dim light conditions. This reflected light causes the "eye shine" that you may have seen when car lights shine on an animal in the dark.

Now study the anterior half of the eyeball (Figs. 7-2 and 7-3B). The choroid and a part of the retina

Eye

95

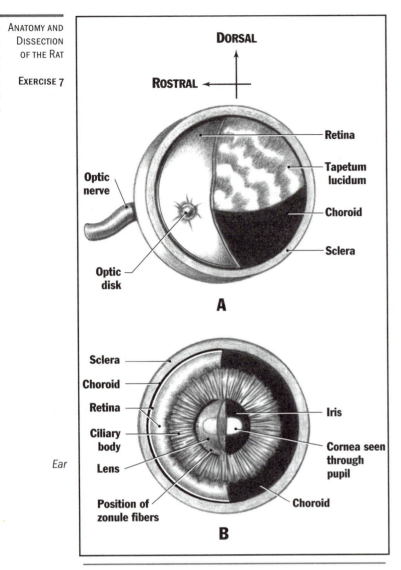

DORSAL

ROSTRAL

Retina

Tapetum lucidum

Choroid

Sclera

Optic nerve

Optic disk

A

Sclera

Choroid

Retina

Ciliary body

Lens

Iris

Cornea seen through pupil

Position of zonule fibers

Choroid

B

Ear

FIGURE 7-3

Views of the inside of the vertically sectioned eyeball
of a sheep. (A) Posterior half after removal of the
vitreous body as seen from the front of the eyeball.
(B) Anterior half after removal of the lens as seen
from the back of the eyeball. A portion of the retina
has been removed in each half to show the underlying
vascular tunic. (Redrawn from Walker, W. F., Jr. and
Homberger, D. G., *Vertebrate Dissection,* 8th ed.,
Philadelphia: Saunders College Publishing, 1992.)

extend toward the **lens.** The lens is held within a ring of
the vascular tunic known as the **ciliary body.** The ciliary
body has a somewhat pleated appearance. Carefully
remove the lens, noticing that it tends to adhere to the
ciliary body because microscopic **zonule fibers** pass from
the ciliary body to the periphery of the lens. The oval
lens of a preserved specimen has lost the clarity and elasticity that it has in life, but try passing it over the words

96

on this page. You may see that they are magnified slightly
because the lens bends, or **refracts,** the light waves.

Now you can see the **iris,** which is an extension of
the vascular tunic in front of the lens, and the opening,
the **pupil,** in the center of the iris. The space between the
iris and the place where the lens you removed was
located is the **posterior chamber;** the space between the
iris and the cornea is the **anterior chamber.** Both of
these chambers are filled with a watery **aqueous humor**
during life. This liquid is secreted by the ciliary body and
maintains a certain intraocular pressure. Surplus aqueous humor is drained through a microscopic canal at the
periphery of the cornea.

Light enters the eyeball through the cornea.
Because the cornea has a refractive index that is considerably greater than that of the surrounding air, the light
waves passing through it are slowed down and refracted
toward the optic axis of the eyeball. The amount of light
that continues through the eyeball is determined by the
diameter of the pupil, which is controlled by radial and
circular smooth muscles in the iris. Light continues
through the lens, which refracts it more and focuses a
sharp, but inverted image, on the retina. The lens resembles the fine adjustment on a microscope or camera,
because it brings the image into sharp focus. However,
unlike the process of focusing a microscope or camera,
focusing in mammals occurs not by moving the lens
back and forth, but by changing its shape. When the eye
is at rest, the intraocular pressure produced by the aqueous humor pushes the wall of the eyeball outward near
the lens. This places the lens under tension, because it is
attached by the zonule fibers to the ciliary body and wall
of the eyeball. The elastic lens is somewhat flattened, and
distant objects are in focus. To focus, or **accommodate,**
for near objects, circular and longitudinal smooth muscle fibers within the ciliary body contract. This muscle
action shortens the distance between the ciliary body
and lens, tension on the lens is reduced, and the elastic
recoil of the lens causes it to bulge slightly. Its increased
thickness increases the refraction of light, and close
objects are brought into focus.

The human eyeball is structurally and functionally
very similar to the one you have been studying. The lens
loses its elasticity with age, and older people often are
not able to focus on near objects without glasses. The
lens may also become cloudy with age, giving rise to a
cataract; if cataracts develop, the lens can be replaced in
a surgical procedure.

B. EAR

The mammalian ear has a dual function: sensing changes
in the body's position (equilibrium) and detecting
sound. It consists of three parts: **inner ear, middle ear,**

and **external ear.** The most conspicuous part of the external ear is the large flap, or **auricle** (Fig. 7-1), which, like an old-fashioned ear trumpet, gathers and focuses the sound waves on the eardrum. The human auricle does not normally move, but you may have noticed that the auricle of a dog or cat is turned and directed toward the source of a sound the animal is interested in. A canal, called the **external acoustic meatus,** extends ventrally and medially from the base of the auricle to the eardrum, or tympanic membrane.

Remove the surrounding tissue from the ventral surface of the meatus, trace it as far as you can underneath the skull, and then open it by making an incision as indicated in Fig. 7-1. The delicate **tympanic membrane,** which is set in vibration by the sound waves, can be seen at the end of the external acoustic meatus. Three very small bones, the **auditory ossicles,** transmit the sound waves across the cavity of the middle ear (**tympanic cavity**) to the **inner ear.** One of the auditory ossicles can often be seen through the translucent membrane. Break the tympanic membrane and open the tympanic cavity. As you have already learned (Exercise 3), this cavity is connected to the pharynx by the **auditory** (Eustachian) **tube** (Fig. 7-4). This passage allows the equalization of pressure on each side of the tympanic membrane, thereby preventing the tympanic membrane from being subjected to tension due to unequal air pressure on both sides of the membrane, and allowing it to vibrate freely.

The inner ear contains the receptor cells for both equilibrium and sound detection. These cells lie within liquid-filled sacs and ducts, which are embedded deep within the otic capsules of the skull (Exercise 1). It is impractical to dissect the inner ear, which is best examined on models or demonstrations, if they are available. The semicircular ducts, utriculus, and sacculus of the inner ear (collectively called the vestibular apparatus) are related to the sense of equilibrium. Hairlike cytoplasmic processes of the receptive cells are displaced by movements of the head, thereby allowing an animal to detect changes in body position and movement. Similar receptive cells in the coiled, snail-like cochlea are activated by pressure waves that are generated in the liquid within the cochlea by sound waves. Sound waves attenuate rapidly in air and must be amplified to set up pressure waves in the denser cochlear liquids. This is accomplished by the difference in size between the relatively large tympanic membrane and the very small opening on the side of the otic capsule through which the auditory ossicles push on the cochlear liquid. Virtually all of the energy impinging on the tympanic membrane is concentrated on this small opening.

The human middle and inner ears are essentially the same as those of the rat.

C. NOSE

Nose

Parts of the nose were studied with the respiratory system. Dissect the nose on your specimen at this time. Cut off the head, remove the lower jaw, and then, using a strong scalpel, make a sagittal section through the skull.

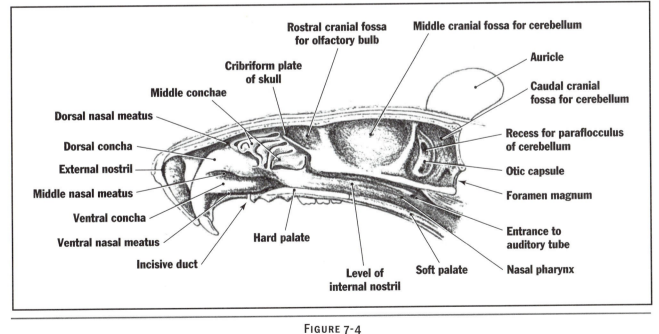

Labels:
Rostral cranial fossa for olfactory bulb
Middle cranial fossa for cerebellum
Cribriform plate of skull
Auricle
Middle conchae
Caudal cranial fossa for cerebellum
Dorsal nasal meatus
Dorsal concha
Recess for paraflocculus of cerebellum
External nostril
Otic capsule
Middle nasal meatus
Foramen magnum
Ventral concha
Ventral nasal meatus
Entrance to auditory tube
Incisive duct
Hard palate
Soft palate
Nasal pharynx
Level of internal nostril

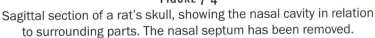

FIGURE 7-4
Sagittal section of a rat's skull, showing the nasal cavity in relation to surrounding parts. The nasal septum has been removed.

A vertical **nasal septum,** composed partly of bone and partly of cartilage, separates the two **nasal cavities.** On the larger section of the skull, pick the septum away and examine the exposed **nasal cavity** (Fig. 7-4). It lies dorsal to the bony **hard palate.** Air enters through the **external nostril** (naris) and leaves through the **internal nostril** (*choana*) to enter the **nasal pharynx** lying dorsal to the fleshy **soft palate.** The surface area within the nasal cavity is increased greatly by a series of bony folds known as the **nasal conchae,** or turbinates. They are covered by a mucous membrane, and their pattern is shown in Fig. 7-4. The air passages between them are the **nasal meatuses,** which converge caudally and enter the nasal pharynx by way of the **internal nostril** (*choana*). The nasal conchae increase the surface area available for olfaction and for conditioning of the respiratory air. As inspired air crosses the nasal conchae, it is warmed and moistened and dust particles are entrapped in mucus secreted by the mucous membrane. As moist air from the lungs is expired, some of the water within it condenses on the mucous membrane and is reabsorbed. The middle, or ethmoid, concha lies just in front of the rostral cranial fossa, which contains the olfactory bulb of the brain. Branches of the olfactory nerve ramify over the conchae and, in particular, the middle concha. The receptive olfactory cells are modified olfactory nerve cells. Human conchae are neither as large nor as complexly folded as those in other mammals, such as the rat. Accordingly, human beings have a poor sense of smell relative to that of most other mammals.

You may see a minute **incisive duct** (Fig. 7-4) extending from the mouth cavity through the palatine fissure of the hard palate (see Fig. 1-5). This duct leads into a small, longitudinal canal that is embedded in the base of the nasal septum and probably will not be seen. This blind (dead-end) canal is the **vomeronasal organ,** which is supplied by a special branch of the olfactory nerve. The vomeronasal organ of many mammals is sensitive to **pheromones,** or chemical markers deposited in the environment by other members of the same species. In particular, it is sensitive to pheromones important in social and sexual interactions between members of a species. Pheromones are used to mark territories, to indicate social status and sexual state, and to convey other social messages. Although this organ starts to develop in the human fetus, anatomists are not certain that it persists as a functional organ in adults. It is absent in aquatic mammals, such as whales and porpoises.

Nose

APPENDIX

GLOSSARY

Glossary of Vertebrate Anatomical Terms

T HIS GLOSSARY IS A BASIC VOCABULARY of anatomical terms that students will encounter in using *Anatomy and Dissection of the Fetal Pig, Anatomy and Dissection of the Rat, Dissection of the Frog,* and many other exercises dealing with vertebrates. Most entries include the term, its pronunciation, its derivation, and its definition.

The pronunciation is given by a simple phonetic spelling in brackets. Stressed syllables are marked by a prime ('); others are separated by hyphens. Long vowels are indicated by the macron (¯); short ones, by the breve (˘).

The classical derivation in parentheses typically includes an abbreviation of the language of the word, the original word in italics, and the meaning of the original word. Usually only the nominative singular is given for Greek and Latin nouns, but the genitive (gen.) sometimes is used if it is closer to the root. Latin adjectives are given only in the masculine form of the nominative singular. Greek and Latin verbs usually are given in the first person present tense because this is closer to the root than the infinitive (but the meaning is given in the infinitive). The present or past participle (pres. p. or p. p.) is given when one of them is closer to the root. Many Greek and Latin words have a combining form that is used when the word is used in combination with other words. Combining forms are indicated by a hyphen before or after the word, for example, L. **Inter-** = between, as in interstitial.

When the English and classical terms are identical, the term is not repeated in the derivation, for example, **Acetabulum** (L. = vinegar cup). When two or more successive terms use the same root, the derivation of the root is given only for the first term. The origin of many repetitive terms is given only under the first entry. For example, **Ligamentum arteriosum** is defined the way this combination of terms is used, but the derivations of ligamentum and arteriosum will be found under **Ligament** and **Artery**. Names of individual muscles that are descriptive of an easily recognized feature of the muscle (its shape or attachments) are not given, but the component parts of less familiar ones are included. For example, the omo-hyoid muscle is not listed, but **Omo-** and **Hyoid** are. Similarly, names of blood vessels and nerves that simply state the organ supplied are omitted, but the organs are usually listed.

This glossary is not exhaustive. The pronunciation, derivation, and meaning of additional terms can be found in unabridged and medical dictionaries.

99

ANATOMY AND
DISSECTION
OF THE RAT

APPENDIX

Glossary

ABD–AOR

Abdomen [ab′dō-men] (L. from *abdo* = to conceal).
The part of the body containing the visceral organs, limited in mammals to the part caudal to the diaphragm.

Abducens nerve [ab-du′senz] (L. leading away, from *ab-* = away from + *duco*, pres. p. *ducens* = leading).
The sixth cranial nerve; carries motor fibers to an extrinsic ocular muscle.

Abduction [ab-dŭk′shŭn] (L. *duco*, p. p. *ductus* = to lead).
Muscle action that carries a body part away from a point of reference, often the midventral line of the body.

Accessory nerve.
The eleventh cranial nerve of amniotes. It carries motor fibers to certain branchiomeric shoulder muscles: the trapezius and sternocleidomastoid muscles.

Acetabulum [as-ĕ-tab′yū-lŭm] (L. = vinegar cup, from *acetum* = vinegar).
The cup-shaped socket in the pelvic girdle that receives the head of the femur.

Achilles tendon.
See **Tendon of Achilles.**

Acromion [ă-krō′mē-on] (Gr. *akron* = tip + *omos* = shoulder).
A process on the scapula with which the clavicle articulates in species with a well-developed clavicle.

Adduction [ă-dŭk′shŭn] (L. *ad-* = toward + *duco*, p.p. *ductus* = to lead).
Muscle action that pulls a body part toward a point of reference, often the midventral line of the body.

Adrenal gland [ă-drē′năl] (L. *ad-* = toward, beside + *ren* = kidney).
An endocrine gland located cranial to the kidney (mammals) or on its ventral surface (frogs). Its medullary hormone helps the sympathetic nervous system adjust the body to stress; its cortical hormones help regulate sexual development and the metabolism of minerals, carbohydrates, and proteins. It is also called the *suprarenal gland.*

Allantois [ă-lan′tō-is] (Gr. *allos* = sausage + *eidos* = appearance).
An extraembryonic membrane of reptiles, birds, and mammals. It develops as an outgrowth of the embryonic hindgut. It accumulates waste products. Its vascularized wall serves as a gas exchange organ in embryonic reptiles and birds, and it contributes to the formation of the placenta in eutherian mammals. Its base develops into the urinary bladder of adult mammals.

Alveolus [al-vē′ō-lŭs] (L. = small cavity).
One of a group of small, thin-walled, and vascularized sacs at the termination of the mammalian respiratory tree where gas exchange occurs.

Amnion [am′nē-on] (Gr. = the membrane around the fetus).
The innermost of the extraembryonic membranes. It envelops the fetus and contains amniotic fluid.

Amniote [am′nē-ōt].
A vertebrate whose embryo has an amnion; a reptile, bird, or mammal.

Amphibian [am-fi′bē-ăn] (Gr. *amphi* = both, double + *bios* = life).
A frog, salamander, or other member of the ancestral class of terrestrial vertebrates. Usually it is aquatic as a larva and terrestrial as an adult.

Ampullary gland [am-pul′lăry] (L. *ampulla* = small vessel).
A small gland associated with the terminal end of the ductus deferens in male rats. It contributes to the seminal fluid.

Anamniote [an-am′nē-ōt] (Gr. *an* = without + *amnion* = the membrane around the fetus).
A vertebrate without an amnion; a fish or amphibian.

Antebrachium [an-te-brā′kē-ŭm] (L. *ante* = before + *brachium* = upper arm).
The forearm.

Anterior.
A direction toward the front or belly surface of a human being; sometimes also used for the head end of a quadruped, but *cranial* is a more appropriate term.

Anterior chamber.
The space within the eyeball between the cornea and iris. It is filled with aqueous humor.

Anterior commissure.
An olfactory commissure within the cerebrum. It is located just rostral to the third ventricle.

Antrum [an′trŭm] (Gr. *antron* = a cave).
An enclosed cavity within an organ, such as the antrum in a secondary follicle in the mammalian ovary.

Anura [an-yūr′a] (Gr. *an* = without + *oura* = tail).
The amphibian order to which frogs and toads belong.

Anus [ā′nŭs] (L. = the seat, anus).
The caudal opening of the digestive tract in a mammal.

Aorta [ā-ōr′tă] (Gr. *aorte* = great artery).
The major artery carrying blood from the heart to the body. It is sometimes called the dorsal aorta to distinguish it from the ventral aorta of a fish, which carries blood from the heart to the gills.

Aortic valve.

A set of three semilunar-shaped folds at the base of the mammalian aorta. It prevents a backflow of blood into the left ventricle.

Aponeurosis [ap-ō-nū-rō′sis] (Gr. *apo* = from + *neuron* = sinew, nerve).

A sheetlike tendon of a muscle.

Appendix [ă-pen′diks] (L. *appendo* = to hang something on).

A dangling extension of an organ, such as the vermiform appendix at the end of the caecum of some mammals.

Aqueduct of Sylvius (*Franciscus Sylvius*, Dutch Anatomist, 1614–1672).

See **Cerebral aqueduct.**

Aqueous humor [ā′kwē-ŭs hyū′mer] (L. *aqua* = water + *humor* = fluid).

A watery liquid in the anterior and posterior chambers of the eyeball. It is secreted by the ciliary body.

Arachnoid [ă-rak′noyd] (Gr. *arachne* = spider + *eidos* = appearance).

One of three meninges of mammals. It is located between the dura mater and pia mater, and is connected to the pia mater by many strands, which give it a "spider web" appearance.

Arbor vitae [ar′bōr vīt′ē] (L. *arbor* = tree + *vita* = life).

The tree-shaped configuration of white matter within the mammalian cerebellum.

Archinephric duct [ar′ki-nef-rik] (Gr. *arche* = origin, beginning + *nephros* = kidney).

The duct, common to many fishes and amphibians, that drains the kidney. It also transports sperm in the male of many vertebrates. It becomes the ductus deferens in a male mammal. It is also called the *mesonephric duct.*

Areola [ă-rē′ō-lă] (L. = small space).

A small space or area, such as the small, round protuberances on the surface of a pig's chorion.

Arrector pili [ă-rek′tōr pi′li] (L. the raisor, from *arrectus* = upright + *pili* = hairs).

One of the small muscles in the skin of mammals that attach onto the hair follicles and raise the hairs.

Artery [ar′ter-ē] (L. *arteria* = artery).

A vessel that carries blood away from the heart. The blood is usually rich in oxygen, but it may be low in oxygen, as, for example, in the pulmonary arteries, which carry blood from the heart to the lungs.

Artiodactyla [ar′ti-ō-dak-til-a] (Gr. *artios* = even + *daktylos* = finger or toe).

The mammalian order that includes hoofed mammals with an even number of toes: pigs, deer, cattle.

Atlas [at′las] (Gr. mythology, a god who supported the Earth upon his shoulders).

The first cervical vertebra that supports the skull.

Atrioventricular valve [ā′trē-ō-ven-trik′yū-lar] (L. *atrium* = entrance hall + *ventriculus* = a small belly).

The valve between an atrium and ventricle of the heart. It prevents the backflow of blood from the ventricle into the atrium. The right atrioventricular valve of mammals has three cusps, or folds, and is also called the *tricuspid valve;* the left one has two cusps and is also called the *bicuspid,* or *mitral, valve.*

Atrium [ā′trē-um].

A chamber, such as the atrium of the heart, that receives blood from the sinus venosus or veins.

Auditory tube [aw′di-tōr-ē] (L. *audio,* p. p. *auditus* = to hear).

The tube that connects the tympanic (middle ear) cavity and the pharynx. It equalizes air pressure on each side of the tympanic membrane. Also called the *Eustachian tube.*

Auricle [aw′ri-kl] (L. *auricula* = little ear).

The external ear flap or pinna; also the ear-shaped lobe of a mammalian atrium.

Autonomic nervous system [aw-tō-nom′ik] (Gr. *autos* = self + *nomos* = rule, law).

An involuntary part of the nervous system (it "rules" itself.) It supplies motor fibers to glands and visceral organs.

Axillary [ak′sil-ār-ē] (L. *axilla* = armpit).

Pertaining to the armpit, e.g., the axillary artery.

Axis [ak′sis] (L. = axle, axis).

The second cervical vertebra of a mammal. Rotation of the head occurs between the atlas and axis.

Axon [ak′son] (Gr. = axle, axis).

The long, slender process of a neuron specialized to conduct nerve impulses a considerable distance.

Azygos vein [az′ī-gos] (Gr. *a* = without + *zygon* = yoke).

An unpaired vein in mammals. It drains most of the intercostal spaces on both sides of the body.

Basophil [bā′sō-fil] (Gr. *baso-* = alkaline + *phileo* = to love, to have an affinity for).

A leukocyte whose cytoplasmic granules take alkaline stains and appear blue; the rarest form of leukocytes.

Biceps [bī′seps] (L. *bi* = two + *ceps* = head).

A structure with two heads, such as the biceps brachii muscle, which is two-headed in human beings.

Bicornuate [bī-kōr′nū-āt] (L. *cornu* = horn).

A structure with two horns, such as a bicornuate uterus.

Bladder.
A membranous sac in which liquid may accumulate, e.g., the urinary bladder.

Blind spot.
See **Optic disk.**

Blood.
The liquid circulating in the blood vessels, consisting of a liquid plasma and cellular elements.

Bone.
The hard skeletal material of vertebrates consisting of an extracellular matrix of collagen fibers. It is mineralized with calcium phosphate crystals by bone forming cells.

Bowman's capsule. (*Sir William Bowman*, British Anatomist, 1816–1832.)
See **Glomerular capsule.**

Brachial [brā′kē-ăl] (L. *brachium* = upper arm).
Pertaining to the upper arm, e.g., brachial artery, coracobrachialis muscle.

Brachial plexus.
The network of nerves that supplies the shoulder and arm.

Brain.
The enlarged, cranial portion of the central nervous system. It is the major integrative center of the nervous system.

Braincase.
The cartilages and bones that encase the brain; also called the *cranium.*

Branchiomeric [brang′kē-o-mēr′ik] (Gr. *branchia* = gill + *meros* = part).
Pertaining to muscles or other structures associated with or derived from the visceral arches and gills.

Broad ligament.
Mesentery in female mammals. It anchors the reproductive tract to the dorsal body wall.

Bronchus [brong′kŭs] (Gr. *bronchos* = windpipe).
A branch of the trachea, which enters the lungs.

Buccal [bŭk′ăl] (L. *bucca* = cheek).
Pertaining to the mouth, e.g., buccal cavity.

Bulbourethral gland [bŭl′bō-yū-rē-thrăl] (L. *bulbus* = a bulbous root + Gr. *ourethra* = urethra).
A gland in male mammals near the base of the penis. It contributes to the seminal fluid. Also called *Cowper's gland.*

Caecum [sē′kŭm] (L. *caecus* = blind).
A pouch at the beginning of the large intestine. It is very long in many herbivores, including the rat, and houses bacteria and protozoa that digest cellulose.

Calcaneus [kal-kā′nē-ŭs] (L. = heel).
The large proximal tarsal bone that forms the "heel bone" in mammals.

Calyx, pl. calyces [kā′liks, kal′i-sēz] (Gr. *kaylx* = cup).
A cuplike compartment, such as one of the renal calyces, or subdivisions of the renal pelvis within the kidney.

Canaliculi [kan-ă-lik′yū-lī] (L. = small canals).
Minute canals in the bone matrix containing the slender processes of bone-forming cells.

Canine tooth [kā′nīn] (L. *canis* = dog).
The large pointed tooth in mammals. It is located caudal to the incisor teeth. In humans, the crown of the canine resembles an incisor, but its root is much larger. It is absent in the rat and enlarged in the pig.

Capillary [kap′i-lār-ē] (L. *capillus* = hair).
One of the minute, thin-walled blood vessels connecting small arteries to small veins. Exchanges between the blood and interstitial fluid occur across capillary walls.

Cardiac [kar′dē-ak] (Gr. *kardia* = heart).
Pertaining to the heart or its vicinity.

Carotid artery [ka-rot′id] (Gr. *karotides* = large neck artery, from *karoo* = to put to sleep, because compressing the artery leads to unconsciousness).
One of the arteries supplying blood to the head.

Carotid body.
A small enlargement at the junction of the external and internal carotid arteries. It is a chemoreceptor monitoring the levels of oxygen and carbon dioxide in the blood.

Carpal [kar′păl] (Gr. *karpos* = wrist).
One of the small bones forming the wrist.

Cartilage [kar′ti-lij] (L. *cartilago* = gristle).
A firm but elastic skeletal tissue whose matrix contains proteoglycan molecules that bind with water. Cartilage occurs in all embryos and in the skeleton of adult terrestrial vertebrates where firmness, smoothness, and flexibility are needed.

Caudal [kaw′dăl] (L. *cauda* = tail).
Pertaining to the tail or to a direction toward the tail in a quadruped.

Cecum.
See **Caecum.**

Cell body.
The part of a nerve cell, or neuron, that contains the nucleus. It does not include the axon and dendrites.

Central canal.
The cavity in the center of the spinal cord. It is continuous with the fourth ventricle in the brain and contains cerebrospinal fluid.

Central nervous system.
The part of the nervous system consisting of the brain and spinal cord.

Centrum.
See **Vertebral body.**

Cephalic [se-fal'ik] (Gr. *kephale* = head).
Pertaining to the head, e.g., brachiocephalic artery.

Cerebellum [ser-ĕ-bel'um] (L. = small brain).
The dorsal part of the metencephalon. It is an important center for muscular coordination and equilibrium.

Cerebral aqueduct [se-rē'brăl] (L. *cerebrum* = brain + *aqua* = water).
The narrow passage within the brain connecting the third and fourth ventricles; also known as the *aqueduct of Sylvius.*

Cerebral hemisphere.
One of two hemispheres that form most of the cerebrum. They are the major integration centers in the brain of mammals.

Cerebral peduncle [pe-dung'kl] (L. *pedunculus* = little foot).
One of a pair of neuronal tracts, which can be seen on the ventral surface of the mammalian brain. It carries impulses from the cerebrum to motor centers in the brain stem and spinal cord.

Cerebrospinal fluid.
A lymphlike fluid that circulates within and around the central nervous system and helps to protect and nourish it.

Cerebrum [se-rē'brŭm].
The major part of the telencephalon. It includes the cerebral hemispheres and a group of deep nuclei.

Cervical [ser'vĭ-kal] (L. *cervix*, gen. *cervicis* = neck).
Pertaining to the neck, e.g., cervical vertebrae.

Cervix [ser'viks].
The neck of an organ, e.g., the cervix of the uterus.

Choana [kō'an-ă] (Gr. *choane* = funnel).
One of the paired openings between the nasal cavity and pharynx; an internal nostril.

Choledochal duct.
See **Common bile duct.**

Chorda tympani [kōr'dă tim-pan'ē] (L. = cord + Gr. *tympanon* = drum).
A branch of the facial nerve crossing the malleus and traversing the tympanic cavity on its way to innervate certain taste buds on the tongue and salivary glands.

Chorion [ko'rē-on] (Gr. = skinlike membrane enclosing the fetus).
The outermost extraembryonic membrane of reptiles, birds, and mammals.

Choroid [kō'royd] (Gr. *chorioeides* = resembling a chorion).
The highly vascularized middle tunic of the eyeball lying between the retina and fibrous tunic.

Choroid plexus [plek'sŭs] (L. *plexus* = network).
Vascular tufts that project from a thin layer of brain tissue and protrude into a ventricle of the brain. They secrete the cerebrospinal fluid.

Chromatophore [krō-mat'ō-fōr] (Gr. *chromo* = color + *phoros* = bearing, from *pherein* = to bear).
A cell in nonmammalian vertebrates that contains pigment granules.

Ciliary body [sil'ē-ar-ē] (L. *cilium* = eyelash).
A part of the vascular tunic of the eyeball attached to the lens. Its muscle fibers regulate accommodation, and its secretory cells produce the aqueous humor.

Clavicle [klav'i-kl] (L. *clavicula* = small key).
The collar bone. It extends between the scapula and sternum in species in which it is well developed.

Cleido- [klī-do'] (Gr. *kleis*, gen. *kleidos* = key, clavicle).
A root referring to the clavicle; used in combination with other terms, e.g., cleidobrachialis muscle.

Clitoris [klit'ō-ris] (Gr. *kleitoris* = hill).
The small erectile organ of female mammals. It corresponds to much of the penis in a male mammal.

Cloaca [klō-ā'kă] (L. = sewer).
A chamber in nonmammalian vertebrates that receives the termination of the digestive, urinary, and genital tracts.

Coagulating gland.
A gland closely associated with the vesicular gland in male rats. It contributes to the seminal fluid.

Coccyx [kok'siks] (Gr. *kokkyx* = cuckoo).
The several fused caudal vertebrae of human beings. As a whole, the coccyx resembles the shape of the bill of a cuckoo. It does not reach the body surface but serves as the attachment site for some muscles.

Cochlea [kok'lĕ-a] (L. = snail shell).
Spiral part of the inner ear of mammals containing the auditory receptors and associated structures.

Coeliac artery [sē'lē-ak] (Gr. *koilia* = belly).
A branch of the aorta that supplies the cranial abdominal viscera, including the spleen, stomach, and liver.

Coelom [sē'lom] (Gr. *koiloma* = a hollow).
A body cavity that is completely lined by serosa, an epithelium of mesodermal origin.

Collagen [kol′lă-jen] (Gr. *kolla* = glue + *genos* = descent).
Minute protein fibers that form most of the extracellular material in connective tissues.

Colliculus [ko-lik′yū-lŭs] (L. = little hill).
One of four small elevations on the dorsal surface of the mammalian mesencephalon, which are centers for certain optic (superior colliculi) and auditory (inferior colliculi) reflexes.

Colon [kō′lon] (Gr. *kolon* = large intestine).
Most of the large intestine. Depending on the species, it extends from the small intestine or caecum to the cloaca or anus.

Columella [kol-yū-mel′a] (L. = small column).
A term frequently used for the rod-shaped stapes of nonmammalian terrestrial vertebrates. See also **Stapes.**

Commissure [kom′i-syūr] (L. *commissura* = a joining together).
A neuronal tract that interconnects structures on the left and right sides of the central nervous system.

Common bile duct.
The principal duct carrying bile to the intestine. It is formed by the confluence of hepatic ducts from the liver and, when present, the cystic duct from the gallbladder.

Concha [kon′kă] (Gr. *konkhe* = seashell).
One of several folds within the mammalian nasal cavity, which increase surface area; also called a turbinate bone.

Condyle [kon′dīl] (Gr. *kondylos* = knuckle).
A rounded articular surface, such as an occipital condyle.

Condyloid process.
The process of a mammalian mandible. It bears the facet for the jaw joint.

Conjunctiva [kon-jūnk-tī′va] (L. *conjunctus* = joined together).
The epithelial layer that covers the surface of, and fuses with, the cornea. It continues over the inner surface of the eyelids.

Connective tissue.
A widespread body tissue characterized by an extensive extracellular matrix containing fibers and cells. It includes collagenous and elastic connective tissue, fat, cartilage, and bone.

Conus arteriosus [kō′năs ar-ter-ē-ō′sas] (L. = cone).
A chamber of the heart in fishes and amphibians into which the single ventricle discharges blood. It contributes to the bases of the pulmonary trunk and aorta in mammals.

Coracoid [kōr′ă-koyd] (Gr. *korax*, gen. *korakos* = crow + *eidos* = appearance).
The bone forming the caudoventral part of the pectoral girdle in nonmammalian terrestrial

vertebrates. It is reduced to a process (coracoid process) resembling a crow's beak in therian mammals.

Cornea [kōr′nē-ă] (L. *corneus* = horny).
The transparent part of the eyeball through which light passes. It is part of the fibrous tunic.

Coronary ligament [kōr′o-nār-ē] (L. *corona* = crown).
Peritoneum that bridges the gap between the liver and the diaphragm in mammals.

Coronary vessels.
Blood vessels that supply the heart musculature. During part of their course, they encircle the heart between the atria and ventricles.

Coronoid process.
The uppermost process of the mammalian mandible to which certain jaw muscles attach.

Corpus [kōr′pŭs] (L. = body).
The main body or part of an organ.

Corpus callosum [ka-lō′sum] (L. *callosus* = hard).
The large commissure interconnecting the two cerebral hemispheres in mammals.

Corpus cavernosum penis [kav-er-nō′sum pē′nis] (L. *caverna* = hollow space + *penis* = tail, penis).
One of a pair of columns of erectile tissue forming much of the penis.

Corpus luteum [lu-te′um] (L. *luteus* = yellow).
A yellowish endocrine gland within the ovary that develops from the ovulated follicle. The principal hormone it produces is progesterone.

Corpus spongiosum penis [spŭn-je′ō-sum] (Gr. *spongia* = sponge).
A column of erectile tissue that surrounds the penile portion of the urethra and forms the glans penis at the tip of the penis.

Cortex [kōr′teks] (L. = bark).
A layer of distinctive tissue on the surface of many organs, e.g., the cerebral cortex.

Costal [kos′tăl] (L. *costa* = rib).
Pertaining to the ribs, e.g., the costal cartilages.

Cowper's gland (*William Cowper*, British anatomist, 1666–1709).
See **Bulbourethral gland.**

Cranial [krā′nē-ăl] (Gr. *kranion* = skull).
Pertaining to the cranium; also a direction toward the head.

Cranium.
The skull, especially the part encasing the brain.

Cremasteric pouch [krē-mas-ter′ik] (Gr. *kremaster* = suspender).
Layers of the body wall that suspend the testis within the scrotum.

Crus, pl. **crura** [krūs, krū′ră] (L. = lower leg).
The lower leg or shin.

Cucullaris muscle [kyū′kū-lar-is] (L. *cucullus* = cap, hood).
A branchiomeric muscle in fishes and amphibians covering the craniodorsal part of the shoulder. It gives rise to the mammalian trapezius and sternocleidomastoid groups of muscles.

Cutaneous [kyū-tā′nē-ŭs] (L. *cutis* = skin).
Pertaining to the skin.

Cystic duct [sis′tik] (Gr. *kystis* = bladder).
The duct of the gallbladder. It joins the hepatic ducts to form the common bile duct.

Decussation [dē-kŭ-sā′shun] (L. *decusso*, p. p. *decussatus* = to divide crosswise in an X).
The crossing of neuronal tracts in the midline of the central nervous system.

Deferent duct.
See **Ductus deferens.**

Deltoid muscle [del′tōyd] (Gr. *deltoeides* = shaped like the letter delta, Δ).
A muscle crossing the lateral surface of the shoulder. It is shaped like the Greek letter *delta* in human beings.

Dendrite [den′drīte] (Gr. *dendron* = tree).
A highly branched and usually short process of a neuron that receives nerve impulses.

Dermis [der′mis] (Gr. *derma* = skin, leather).
The deeper layer of the skin. It is composed of dense connective tissue and develops from embryonic mesoderm.

Diaphragm [dī′ă-fram] (Gr. *dia* = through, across + *phragma* = partition, wall).
A mostly muscular partition between the thoracic and abdominal cavities in a mammal.

Diencephalon [dī-en-sef′ă-lon] (Gr. *enkephalos* = brain).
The brain region between the telencephalon and mesencephalon. It includes the epithalamus, thalamus, and hypothalamus.

Digit [dij′it] (L. *digitus* = finger).
A finger or toe.

Distal [dis′tăl] (L. *distalis* = situated away from the center).
The end of a structure most distant from its origin.

Dorsal [dōr′săl] (L. *dorsalis*, from *dorsum* = back).
A direction toward the surface of the back of a quadruped.

Duct [dŭkt] (L. *duco*, p. p. *ductus* = to lead).
A small, tubular passage carrying products away from an organ.

Ductus arteriosus [dŭk′tŭs ar-tēr′ē-ō-sŭs].
A connection in the fetal mammal between the pulmonary trunk and aorta, that permits much blood to bypass the embryonic lungs. It atrophies after birth and forms the ligamentum arteriosum.

Ductus deferens [def′er-enz] (L. *defero*, pres. p. *deferens* = to carry away).
The sperm duct of mammals and other amniotes; also called *vas deferens.*

Ductus venosus [vē-nō′sŭs].
A blood vessel in the liver of a fetal mammal. It permits much of the blood returning from the placenta in the umbilical veins to bypass the hepatic sinusoids and to enter the caudal vena cava directly.

Duodenum [dū-ō-dē′nŭm] (L. *intestinum duodenum digitorum, duodeni* = twelve fingers each).
The first part of the small intestine. In human beings, it is about 12 finger-breadths long.

Dura mater [dū-ră mā′ter] (L. = *durus* = hard + *mater* = mother).
The tough outer meninx surrounding the mammalian central nervous system.

Efferent ductules [ef′er-ent] (L. *ex* = out, away from + *fero*, pres. p. *ferens* = to carry).
Minute ducts in amphibians that carry sperm cells from the testis to the cranial tubules of the kidney; in amniotes, modified kidney tubules that lie in the head of the epididymis and transport sperm cells. These ducts are also called *vasa efferentia.*

Embryo [em′brē-ō] (Gr. *embryon* = embryo, from *en* = in + *bryo* = to swell).
An early stage in development of an individual. It is dependent for energy and nutrients from stored material within its egg or from the mother; i.e., embryos are not able to live on their own.

Endocrine gland [en′dō-krin] (Gr. *endon* = within + *krino* = to separate).
A ductless gland that discharges its secretion (a hormone) into the blood.

Eosinophil [ē-ō-sin′ō-fil] (Gr. *eos* = dawn + *philos* = fond).
A leukocyte whose cytoplasmic granules stain with eosin (an acid dye) and appear reddish.

Epaxial [ep-ak′sē-ăl] (Gr. *epi* = upon, above + *axon* = axle, axis).
Pertaining to those muscles and other organs that lie above or beside the dorsal half of the vertebral column.

Ependymal epithelium [ep-en′di-măl] (Gr. *ependyma* = garment).
The epithelial layer that lines the central canal of the spinal cord and the cavities in the brain.

Epidermis [ep-i-derm′is] (Gr. *epi* = upon + *derma* = skin).
The superficial layer of the skin. It is composed of a stratified, squamous epithelium whose outer layers are keratinized in terrestrial vertebrates. It is derived from the embryonic ectoderm.

Epididymis [ep-i-did′i-mis] (Gr. *didymoi* = testis).
A band-shaped group of tubules and a coiled duct, lying on the testis of mammals and in which sperm cells are stored. It evolved from a part of the kidney and kidney duct of ancestral amphibians and fishes.

Epiglottis [ep-i-glot′is] (Gr. *glottis* = entrance to the windpipe).
The flap of fibrocartilage that helps deflect food around the glottis of mammals during swallowing.

Epiphysis [e-pif′i-sis] (Gr. *physis* = growth).
(1) The end of a long bone in a young mammal. (2) A thin outgrowth of the epithalamus in some fishes and amphibians, the distal end of which is sensitive to changes in light conditions. It corresponds to the mammalian pineal gland.

Epithalamus [ep′i-thal′ă-mŭs] (Gr. *thalamos* = inner chamber).
The roof of the diencephalon lying above the thalamus. Part of it is an olfactory center.

Epithelium [ep-i-thē′lē-um] (Gr. *thele* = delicate skin).
A tissue that is composed of tightly packed cells and covers body surfaces and lines cavities, including those of blood vessels and ducts.

Epitrichium [ep-i-trik′ē-ŭm] (Gr. *trichion* = small hair).
A layer of epithelium that lies upon the developing hairs in a mammal fetus.

Erectile tissue.
A tissue in an organ containing cavernous vascular spaces that may fill with blood and swell.

Esophagus [ē-sof′ă-gŭs] (Gr. *oisophagos* = gullet).
The part of the digestive tract between the pharynx and the stomach.

Eustachian tube (*Bartolommeo Eustachio,* Italian anatomist, 1520-1574).
See **Auditory tube.**

Exocrine gland [ek′sō-krin] (Gr. *ex* = out + *krino* = to separate).
A gland whose secretion is discharged through a duct onto a surface or into a cavity.

Extension [eks-ten′shŭn] (L. *extendo* = to stretch out).
A movement that carries a distal limb segment away from the next proximal segment, retracts a limb at the shoulder or hip, or moves the head or a part of the trunk toward the middorsal line.

External acoustic meatus [ă-kūs′tik mē-ā′tŭs] (Gr. *akoustikos* = pertaining to hearing + L. *meatus* = passage).
The external ear canal of amniotes extending from the body surface to the tympanic membrane.

External nostril.
See **Naris.**

Extrinsic [eks-trin′sik] (L. *extrinsecus* = from without).
A structure that originates from another structure or organ (e.g., extrinsic ocular muscles).

Extrinsic ocular muscles [ok′yū-lăr].
The group of small, strap-shaped muscles that extend from the wall of the orbit to the eyeball.

Facial [fā′shăl] (L. *facies* = face).
Pertaining to the face.

Facial nerve.
The seventh cranial nerve; innervates facial and other muscles associated with the second visceral arch and its derivatives, some salivary glands, and taste receptors on the front of the tongue.

Fallopian tube (*Gabriele Fallopio,* Italian anatomist, 1523-1562).
See **Uterine tube.**

Falx cerebri [falks se-rē′bri] (L. sickle + *cerebrum* = brain)
A sickle-shaped fold of dura mater in the longitudinal fissure between the two cerebral hemispheres.

Fascia [fash′ē-ă] (L. = band).
Sheets of connective tissue that lie beneath the skin (e.g., superficial fascia) or ensheathe groups of muscles (e.g., deep fascia) or organs.

Femur [fē′mŭr] (L. = thigh).
The thigh or bone within the thigh.

Fenestra [fe-nes′tră] (L. = window).
A windowlike opening in an organ.

Fenestra cochleae [kok′lĕ-ē] (L. *cochlea* = snail shell).
A small opening on the medial wall of the tympanic cavity of a mammal from which pressure waves, which have traveled through the inner ear, are released. It is often called the round window.

Fenestra vestibuli [ves-ti′būl-ē] (L. *vestibulum* = antechamber).
A small opening on the medial wall of the tympanic cavity into which the inner end of the stapes fits and propagates pressure waves into the inner ear. It is also called the *oval window.*

Fetus [fē′tŭs] (L. = offspring).
The unborn young of a mammal after it has nearly assumed the appearance it will have at birth (after the eighth week of gestation in human beings).

Fibroblast [fi′brō-blast] (L. *fibra* = fiber + Gr. *blastos* = bud).
An elongated and branching connective tissue cell that produces the intercellular matrix, including collagen fibers.

Fibrous tunic [tū′nik] (L. *tunica* = coat).
The tough connective tissue forming the outer wall of the eyeball; divided into the cornea and sclera.

Fibula [fib′yū-la] (L. = buckle, pin).
The slender bone on the lateral surface of the lower leg.

Fissure [fish′ūr] (L. *fissura* = cleft).
A deep groove or cleft in an organ, such as the brain or skull.

Flexion [flek′shŭn] (L. *flexio* = the bending).
A movement that brings a distal limb segment toward the next proximal one, advances a limb at the shoulder or hip, or bends the head or a part of the trunk toward the midventral line of the body.

Folia [fō′lē-ă] (L. = leaves).
Leaflike folds in an organ, such as those in the cerebellar hemispheres.

Foramen, pl. **foramina** [fō-rā′men, fō-ram′i-nă] (L. = opening).
An opening in an organ.

Foramen magnum [mag′nŭm] (L. *magnus* = large).
The opening in the base of the skull through which the spinal cord passes.

Foramen of Monro (*Alexander Monro, Sr.,* Scottish anatomist, 1697–1767).
See **Interventricular foramen.**

Foramen ovale [ō-vāl′ē] (L. *ovalis* = oval).
A valved opening in the interatrial septum of a fetal mammal. It permits much of the blood (chiefly blood from the placenta and caudal part of the body) to pass from the right to the left atrium, thus, bypassing the lungs. It closes at birth and becomes the adult fossa ovalis.

Forebrain.
See **Prosencephalon.**

Fossa [fos′ă] (L. = ditch).
A shallow groove or depression in an organ.

Fossa ovalis [ō-vah′lis] (L. *ovalis* = oval).
An oval depression in the medial wall of the right atrium of an adult mammal. It is a vestige of the fetal foramen ovale.

Frontal [frŭn′tăl] (L. *frons,* gen. *frontis* = forehead).
(1) Pertaining to the forehead. (2) A plane of the body that passes through the frontal suture between the frontal and parietal bones; i.e., a median longitudinal plane passing from left to right.

Gallbladder [gawl′blad-er] (Old English *galla* = bile).
A small sac attached to the liver in most vertebrates. It stores and concentrates bile. The gallbladder is absent in rats.

Gametes [gam′ētz] (Gr. *gamete* = wife, *gametes* = husband).
The haploid germ cells, eggs and sperm.

Ganglion [gang′glē-on] (Gr. = small tumor, swelling).
A group of neuronal cell bodies that is a part of the peripheral nervous system in vertebrates.

Gastric [gas′trik] (Gr. *gaster* = stomach).
Pertaining to the stomach.

Gemelli muscles [jē-mel′lĭ] (L. = small twins).
Two small muscles situated deeply on the lateral surface of the hip in many mammals. They are fused with the obturatorius muscle in the pig.

Genio- [jē-ni′ō] (Gr. *geneion* = chin).
A combining term pertaining to the chin, as in the genioglossus muscle.

Girdles.
The skeletal elements in the body wall of vertebrates, that support the appendages.

Gland (L. *glans* = acorn).
A group of secretory cells. Exocrine glands discharge their secretion by a duct onto the body surface or into a cavity. Ductless endocrine glands discharge their secretion into the blood.

Glans clitoridis [glanz kli-tō′ri-dis] (Gr. *kleitoris* = hill).
The small mass of erectile tissue at the distal end of the clitoris in a female mammal.

Glans penis [pē′nis] (L. *penis* = tail, penis).
The bulbous, distal end of the penis in a mammal; part of the corpus spongiosum penis.

Glenoid fossa [glen′oyd] (Gr. *glene* = socket + *eidos* = resembling).
The socket that is located on the pectoral girdle of terrestrial vertebrates and receives the head of the humerus.

Glomerular capsule [glō-mār′yū-lăr] (L. *glomerulus* = little ball of yarn).
The thin-walled, expanded proximal end of a renal tubule. It surrounds a tuft of capillaries, the glomerulus. It is also called *Bowman's capsule.*

Glomerulus [glō-mār′yū-lŭs] (L. = little ball of yarn).
A dense network of capillaries that is surrounded by the glomerular capsule at the proximal end of a kidney tubule.

Glossopharyngeal nerve [glos′ō-fă-rin-jē-ăl] (Gr. *glossa* = tongue + *pharynx* = throat).
The ninth cranial nerve. It carries motor fibers to pharyngeal muscles derived from the third visceral arch, parasympathetic fibers to certain salivary glands, and returns sensory fibers from taste buds on part of the tongue and from the part of the pharynx near the base of the tongue.

Glottis [glot′is] (Gr. = opening of the windpipe).
A slitlike opening in the center of the larynx, including the space between the vocal cords.

Gluteal [glū′tē-ăl] (Gr. *gloutos* = buttock).
Pertaining to the buttocks, e.g., gluteal muscles.

Goblet cells.
Goblet-shaped, mucus-producing cells associated with the epithelial lining of the stomach, intestine, and upper respiratory tract.

Gonads [gō′nadz] (Gr. *gone* = seed).
A collective term for the testes and ovaries.

ANATOMY AND
DISSECTION
OF THE RAT

APPENDIX

Glossary

GRA–ILE

Graafian follicle [grä′fē-ăn] (*Reijnier de Graaf,* Dutch anatomist, 1641–1673).
  See **Tertiary follicle.**

Gray matter.
  Tissue in the central nervous system consisting primarily of cell bodies of neurons and unmyelinated nerve fibers.

Gubernaculum [gū′ber-nak′yū-lŭm] (L. = small rudder).
  A cord of connective tissue that lies at the caudal end of the testis and guides the descent of the testis into the scrotum.

Gyrus [jī′rŭs] (Gr. *gyros* = circle).
  One of the folds on the surface of a cerebral hemisphere.

Hair.
  A filamentous epidermal structure in the skin of mammals. It consists primarily of keratinized cells. En masse, it helps to thermally insulate the body.

Hallux [hal′ŭks] (Gr. = big toe).
  The first or most medial digit of the foot.

Harderian gland (*Johann Harder,* Swiss anatomist, 1656–1711).
  A tear gland present in some mammals. It is located rostral and ventral to the eyeball. Also called *accessory lacrimal gland.*

Hard palate.
  The part of the secondary palate that is supported by a horizontal shelf of bone in the mammalian skull and separates the oral from the nasal cavities.

Haversian system (*Clopton Havers,* British anatomist, 1650–1702).
  See **Osteon.**

Heart.
  The hollow muscular organ that pumps blood through the body.

Hemoglobin [hē-mō-glō′bin] (Gr. *haima* = blood + L. *globus* = globe).
  Iron-containing, globular molecules that fill the red blood cells and bond reversibly with oxygen and some carbon dioxide.

Hepatic [he-pat′ik] (Gr. *hepar* = liver).
  Pertaining to the liver.

Hepatic duct.
  One of the ducts that drain the liver. Hepatic ducts join a cystic duct from the gallbladder to form the common bile duct in animals that possess a gallbladder.

Hepatic portal system.
  A system of veins that drain the capillaries of the abdominal digestive organs and spleen and that lead to the sinusoids within the liver.

Hepatic vein.
  One of the veins that drains the hepatic sinusoids and, in mammals, leads to the caudal vena cava.

Hindbrain.
  See **Rhombencephalon.**

Hyaloid artery [hī′ă-loyd] (Gr. *hyolos* = glass + *eidos* = appearance).
  The embryonic artery that passes through the vitreous body of the eyeball to supply the developing lens. It disappears before birth.

Hyobranchial apparatus [hī′ō-brang′kē-ăl] (Gr. *hyo* = a combining form referring to structures associated with the second visceral, or hyoid, arch + *branchia* = gill).
  The complex of cartilages and bones that is derived from the hyoid and other visceral arches and which supports the tongue and the floor of the mouth and pharynx in nonmammalian terrestrial vertebrates. It is also called the hyoid apparatus in mammals.

Hyoid [hī′oyd] (Gr. *hyoeides* = resembling the letter *upsilon* = U-shaped).
  (1) Pertaining to structures associated with the second visceral, or hyoid, arch. (2) The mammalian bone that is embedded in and supports the base of the tongue.

Hyoid apparatus.
  See **Hyobranchial apparatus.**

Hypaxial [hī-pak′sē-ăl] (Gr. *hypo* = under + *axon* = axle or axis).
  Pertaining to those muscles and other structures that lie in the body wall ventral to the dorsal half of the vertebral axis.

Hypobranchial [hī-pō-brang′kē-ăl] (Gr. *branchia* = gill).
  Pertaining to muscles and other structures derived from structures that are located ventral to the gills in fishes.

Hypoglossal nerve [hī-pō-glos′ăl] (Gr. *glossa* = tongue).
  The twelfth cranial nerve of amniotes. It carries motor fibers to the muscles of the tongue.

Hypophysis [hī-pof′i-sis] (Gr. *physis* = growth).
  An endocrine gland that is attached to the ventral surface of the hypothalamus. It secretes many hormones that regulate a variety of physiological processes and other endocrine glands. Also called the pituitary gland.

Hypothalamus [hī-pō-thal′ă-mŭs] (Gr. *thalamus* = inner chamber).
  The ventral part of the diencephalon. It lies ventral to the thalamus and is an important center for the integration of visceral functions.

Ileum [il′ē-ŭm] (L. = small intestine).
  The caudal half of the small intestine of mammals. In human beings it is the part of the small intestine that lies against the ilium.

Iliac [il′ē-ak] (L. *ilium* = groin, flank).
 Pertaining to a structure near the groin or flank, e.g., iliac artery.

Ilium [il′ē-ŭm] (L. = groin or flank).
 The bone forming the dorsal part of the pelvic girdle of terrestrial vertebrates. It articulates with the sacrum.

Incisor tooth [in-ṣi′zŏr] (L. = the cutter, from *incido,* p. p. *incisum* = to cut into).
 One of the front teeth of mammals lying rostral to the canine tooth. It is used for cutting and cropping and is very large in rodents.

Incus [ing′kŭs] (L. = anvil).
 The anvil-shaped, middle auditory ossicle of mammals.

Inferior [in-fē′rē-or] (L. = lower).
 A direction toward the feet of a human being.

Infundibulum [in-fŭn-dib′yŭ-lŭm] (L. = little funnel).
 A funnel-shaped structure, such as the entrance into the oviduct or uterine tube.

Inguinal [ing′gwi-năl] (L. *inguen,* gen. *inguinis* = groin).
 Pertaining to structures in or near the groin.

Inguinal canal.
 A passage through the muscular and fascial layers of the body wall, leading from the abdominal cavity to the vaginal cavity of the scrotum. Ducts, blood vessels, and nerves to and from the testis travel through this passage.

Inner ear.
 The portion of the ear that lies within the otic capsule of the skull and contains the receptive cells for equilibrium and hearing.

Insertion of a muscle.
 The point of attachment of a muscle that moves the greater distance when the muscle contracts. It is usually the distal end of a limb muscle.

Integument [in-teg′yŭ-ment] (L. *integumentum* = covering).
 The body covering consisting of the epidermis and the dermis. It often includes many accessory structures: glands, hair, feathers, scales. Also called skin.

Internal nostril.
 See **Choana.**

Interneuron [in′ter-nū-ron] (L. *inter* = between + Gr. *neuron* = nerve, sinew).
 A neuron that lies within the central nervous system and connects sensory and motor neurons or other interneurons. It is responsible for most of the integrative activity of the nervous system.

Interstitial cells [in-ter-stish′al] (L. *interstitium* = a small space between).
 Groups of endocrine cells that lie between the seminiferous tubules of the testis and secrete the male sex hormone, testosterone.

Interstitial fluid.
 A lymphlike fluid that lies in the minute spaces between the cells of the body.

Interthalamic adhesion [in-ter-tha-lam′ik] (Gr. *thalamos* = inner chamber).
 The part of the mammalian thalamus that crosses the midline within the third ventricle. Also called the massa intermedia.

Interventricular foramen [in-ter-ven-trik′yŭ-lar fō-ra′men] (L. *ventriculus* = belly).
 A foramen leading from one of the lateral ventricles of the brain to the third ventricle. Also called the *foramen of Monro.*

Intestine [in-test′tin] (L. *intestinus* = intestine).
 The portion of the digestive tract between the stomach and anus or cloaca. It is the site of most digestion and absorption.

Intrinsic [in-trin′sik] (L. *intrinsecus* = on the inside).
 A structure that is an inherent part of an organ, e.g., the ciliary muscles of the eyeball.

Iris [ī′ris] (Gr. = rainbow).
 The pigmented part of the vascular tunic of the eyeball. It surrounds the pupil.

Ischiadic, ischiatic [is-ke-ad′ (at) -ik] (Gr. *ischion* = hip).
 Pertaining to a structure near the hip, e.g., the ischiadic nerve.

Ischium [is′kē-ŭm].
 The bone forming the caudoventral part of the pelvic girdle.

Islets of Langerhans (*Paul Langerhans,* German physician, 1847–1888).
 See **Pancreatic islets.**

Jacobson's organ (*Ludwig L. Jacobson,* Danish surgeon and anatomist, 1783–1843).
 See **Vomeronasal organ.**

Jejunum [jĕ-jū′nŭm] (L. *jejunus* = empty).
 Approximately the first half of the mammalian postduodenal small intestine. It is usually found to be empty in autopsies.

Jugular vein [jug′yŭ-lar] (L. *jugulum* = throat).
 One of the major veins in the neck of mammals. It drains the head.

Keratin [ker′ă-tin] (Gr. *keras* = horn).
 A horny protein synthesized by the epidermal cells of terrestrial vertebrates.

Kidney [kid′nē] (Middle English *kindenei* = kidney).
 The organ that removes waste products, especially nitrogenous wastes, from the blood and produces urine.

Kidney tubule.
> See **Renal tubule.**

Kneecap.
> See **Patella.**

Labia [lā′bē-ă] (L. = lips).
> Liplike structures, e.g., the genital labia of a female mammal.

Lacrimal [lak′ri-măl] (L. *lacrima* = tear).
> Pertaining to structures involved with the production and transport of tears (e.g., lacrimal gland) or to structures located near these structures (e.g., lacrimal bone).

Lacuna [lă-kū′nă] (L. = a small hollow or space).
> A small cavity, such as one of many in the bone matrix, in which a bone cell (osteocyte) lodges.

Lamella [lă-mel′ă] (L. = small plate, layer).
> A thin plate or layer of tissue, such as the layers of matrix in bone.

Laryngotracheal chamber.
> A chamber in the respiratory tract of amphibians into which the glottis leads and from which the lungs emerge. It is comparable to the larynx and trachea of mammals.

Larynx [lar′ingks] (Gr. = larynx).
> The part of the respiratory tract of amniotes between the pharynx and trachea. It contains the vocal cords in mammals.

Lateral [lat′er-ăl] (L. *latus*, gen. *lateris* = side).
> Pertaining to the side of the body.

Lens [lenz] (L. = lentil).
> A refractive transparent body near the front of the eyeball. It is responsible for accommodation or focusing by changing its shape (mammals) or by moving toward or away from the retina (frogs).

Leukocyte [lū′kō′sīt] (Gr. *leukos* = clear, white + *kytos* = cell).
> Any of several different types of white blood cells.

Lienic [lī′ě-nik] (L. *lien* = spleen).
> Pertaining to the spleen, e.g., lienogastric artery.

Ligament [lig′ă-ment] (L. *ligamentum* = band, bandage).
> (1) A band of dense connective tissue extending between certain structures, usually bones. (2) A mesentery extending between certain visceral organs.

Ligamentum arteriosum [lig′ă-men-tum ar-tēr′ē-ō-sum].
> A band of dense connective tissue extending from the pulmonary trunk to the aorta in adult mammals. It is a remnant of the embryonic ductus arteriosus.

Linea alba [lin′ē-ă al′ba] (L. = white line).
> A whitish strand of connective tissue in the midventral abdominal wall to which abdominal muscles attach.

Limbic system [lim′bik] (L. *limbus* = border).
> A brain region that encircles the diencephalon and leads to the hypothalamus. It is important in regulating many behaviors related to species survival, including feeding and reproduction.

Lingual [ling′gwăl] (L. *lingua* = tongue).
> Pertaining to the tongue.

Liver [liv′er] (Old English *lifer* = liver).
> The large organ in the cranial part of the abdominal cavity. It secretes bile and plays a vital role in many metabolic processes, including processing many substances brought to it by the hepatic portal vein system, disassembling senescent red blood cells, detoxifying substances, and synthesizing many plasma proteins.

Lumbar [lŭm′bar] (L. *lumbus* = loin).
> Pertaining to structures in the back between the thorax and pelvis, such as the lumbar vertebrae.

Lung [lŭng] (Old English *lungen* = lung).
> One of the paired organs of terrestrial vertebrates in which gases are exchanged between the air and blood. It develops as an outgrowth from the floor of the pharynx.

Lymph [limf] (L. *lympha* = clear water).
> A clear liquid that is derived from the interstitial fluid and flows in the lymphatic vessels. It lacks the red blood cells and most of the plasma proteins found in blood.

Lymph heart.
> A pulsating part of lymphatic vessels of some amphibians and reptiles. It helps return lymph to the veins.

Lymph node.
> Nodules of lymphatic tissue along the course of the lymphatic vessels of mammals. It is the site for the multiplication of many lymphocytes, the phagocytosis of many foreign particles, and the initiation of certain immune response.

Lymphocyte [lim′fō-sīt] (Gr. *kytos* = cell).
> A leukocyte with a large, round nucleus and very little cytoplasm. Lymphocytes develop in lymph nodes, spleen, thymus, and other lymphoid tissues, and participate in immune responses.

Macrophage [mak′rō-fāj] (Gr. *makros* = large + *phagein* = to eat).
> Large cell in the tissues. It phagocytoses foreign particles and participates in immune responses.

Malleus [mal′ē-ŭs] (L. = hammer).
> The outermost of three auditory ossicles of a mammal. It transmits sound waves across the tympanic cavity; it is shaped like a hammer.

Mammal [mam'ăl] (L. *mamma* = breast).
A member of the vertebrate class characterized by hair and mammary glands.

Mammary gland [mam'ă-rē].
The milk-secreting, cutaneous gland that characterizes female mammals.

Mammary papilla [pă-pil'ă] (L. = nipple, pimple).
The nipple, or teat, of a mammary gland.

Mandibular cartilage [man-dib'yū-lăr] (L. *mandibula* = lower jaw).
A cartilaginous core of the bony lower jaw of many fishes, amphibians, and reptiles. It represents the ventral half of the first visceral arch of ancestral fishes.

Mandibular gland.
A mammalian salivary gland usually located deep to the caudoventral angle of the lower jaw.

Mandibular ramus [rā'mŭs] (L. = a branch).
The vertical part of the mandible caudal to the teeth.

Massa intermedia.
See **Interthalamic adhesion.**

Masseter muscle [mă-sē'ter] (Gr. *masseter* = chewer).
A large mammalian muscle that helps to close the jaws. It extends from the zygomatic arch to the mandible.

Mastoid process [mas'toyd] (Gr. *mastos* = breast + *eidos* = appearance).
A large, rounded process on the base of the mammalian skull located caudal to the external acoustic meatus; it is a point of attachment for certain neck muscles.

Meatus [mē-ā'tŭs] (L. = a passage).
A passage, such as the external acoustic meatus, that leads from the head surface to the tympanic membrane.

Meconium [mē-kō'nē-um] (Gr. *mekonion* = poppy juice).
Bile-stained debris in the fetal digestive tract. It is discharged shortly after birth.

Medial [mē'dē-al] (L. *medialis* = middle).
A direction toward the middle of the body.

Median [mē'dē-an] (L. *medianus* = middle).
Lying in the midline of the body.

Mediastinum [me'dē-as-tī-nŭm] (L. = middle septum).
The space in the mammalian thorax between the two pleural cavities. It contains the aorta, esophagus, pericardial cavity, heart, and the venae cavae.

Medulla [me-dūl'ă] (L. = core, marrow).
The central part, or core, of certain organs, as opposed to their surface region, or cortex.

Medulla oblongata [ob-long-gah'tă].
The myelencephalon, or caudal region of the brain, which is continuous with the spinal cord.

Melanin [mel'ă-nin] (Gr. *melas* = black).
A black pigment often contained in cells known as melanophores.

Meninx, pl. **Meninges** [mē'ningks, mĕ-nin'jēz] (Gr. = membrane).
A connective tissue membrane that surrounds the central nervous system.

Mesencephalon [mez-en-sef'ă-lon] (Gr. *mesos* = middle + *enkephalos* = brain).
The middle, or third, of five brain regions. It lies between the diencephalon and metencephalon and includes the colliculi (mammals) or optic lobes (frogs); also called the midbrain.

Mesentery [mez'en-ter-ē] (Gr. *enteron* = intestine).
(1) Any membrane-like double layer of serosa that extends from the body wall to visceral organs, or between visceral organs. (2) The particular membrane suspending the intestine.

Mesonephric duct.
See **Archinephric duct.**

Mesorchium [mez-ōr'kē-ŭm] (Gr. *orchis* = testis).
The mesentery suspending the testis.

Mesovarium [mez'ō-va'rē-um] (L. *ovarium* = ovary).
The mesentery suspending the ovary.

Metacarpal [met'ă-kar'păl] (Gr. *meta* = after + *karpos* = wrist).
One of the long bones in the palm of the hand located between the carpals and phalanges.

Metatarsal [met'ă-tar'săl] (Gr. *tarsos* = ankle).
One of the long bones of the foot located between the tarsals and phalanges.

Metencephalon [met'en-sef'ă-lon] (Gr. *enkephalos* = brain).
The fourth of the five brain regions. It lies between the mesencephalon and myelencephalon and includes the cerebellum and (in mammals) the pons.

Midbrain.
See **Mesencephalon.**

Middle ear.
The portion of the ear of terrestrial vertebrates that usually contains the tympanic cavity and one or three auditory ossicles. Airborne sound waves are transmitted from the body surface (usually from a tympanic membrane) across the middle ear cavity via the auditory ossicle (or ossicles) to the inner ear.

Middle ear cavity.
See **Tympanic cavity.**

Molar tooth [mō'lăr] (L. *molaris* = millstone).
One of the caudal teeth of mammals usually adapted for crushing or grinding food.

ANATOMY AND
DISSECTION
OF THE RAT

APPENDIX

Glossary

MON-OCC

Monocyte [mon′ō-sīt] (Gr. *monos* = single + *kytos* = cell).
A leukocyte with a kidney-shaped nucleus. It is a precursor of a tissue macrophage.

Mouth [mowth] (Old English *muth* = mouth).
The cranial opening of the digestive tract. Also used for the oral cavity into which this opening leads.

Mucosa [myū-kō-să] (L. *mucosus* = slimy, mucous).
The lining of the digestive tract and many other hollow visceral organs. It consists of an epithelium, connective tissue, and sometimes a thin layer of smooth muscle.

Mucus [myū′kŭs] (L. = slime).
A slimy material secreted by some epithelial cells. It is rich in the glycoprotein mucin.

Muscle [mŭs-ĕl] (L. *musculus* = muscle, a little mouse).
(1) A contractile tissue responsible for most of the movements of the body or its parts. (2) A discrete group of muscle fibers with a common origin and insertion.

Myelencephalon [mī′el-en-sef′ă-lon] (Gr. *myelos* = core, spinal cord + *enkephalos* = brain).
The most caudal of the five regions of the brain. It consists of the medulla oblongata.

Mylohyoid [mī′lō-hī′ōyd] (Gr. *myle* = a mill + *hyoeides* = U-shaped).
A nearly transverse sheet of muscle extending between the two mandibles and the hyoid bone.

Myo- [mī′ō] (Gr. *mys*, gen. *myos* = muscle).
A prefix meaning muscle or musclelike.

Myoepithelium [mī′ō-ep-thē-lē-um].
Specialized epithelial cells containing contractile elements. Some surround the secretory cells of sweat glands and help to discharge sweat.

Myofibrils [mī-ō-fī′brilz] (L. *fibrilla* = a minute fiber).
Minute, longitudinal, contractile fibrils within a muscle fiber. They are barely visible with a light microscope and are composed of myofilaments.

Myofilaments [mī′ō-fil′ă-mentz].
Ultramicroscopic filaments of actin and myosin whose interactions are the basis of muscle contraction.

Naris, pl. **nares** [nā′ris, nā′res] (L. = nostril).
An opening from the outside into the nasal cavity; an external nostril.

Nasal [nā′zăl] (L. *nasus* = nose).
Pertaining to the nose, e.g., nasal bone, nasal cavity.

Nasal conchae [kon′kē] (L. = shells).
Folds within the nasal cavity of mammals. They increase the surface area.

Nasal meatus (L. = a passage).
An air passage in the mammalian nasal cavity between the nasal conchae, or between the conchae and nasal septum.

Nephron [nef′ron] (Gr. *nephros* = kidney).
The structural and functional unit of the kidney. It consists of a glomerulus and a renal tubule.

Nerve [nerv] (L. *nervus* = nerve, sinew).
A bundle of neuronal processes (axons) and their investing connective tissues. It extends between the central nervous system and peripheral organs, and is part of the peripheral nervous system.

Neural arch.
See **Vertebral arch.**

Neuron [nūr′on] (Gr. *neuron* = nerve, sinew).
A nerve cell. It is the structural and functional unit of the nervous system. It consists of dendrites, a cell body, and an axon.

Neuron tract.
A bundle of neuronal processes within the spinal cord or brain.

Neutrophile [nū′trō-fil] (L. *neutro* = neither + Gr. *philos* = affinity for).
A leukocyte with small cytoplasmic granules that stain only lightly with basic and acidic dyes. Neutrophiles are phagocytic and are the most abundant leukocytes.

Nictitating membrane [nik′ti-tāt-ing] (L. *nicto* = to beckon, wink).
A membrane in many terrestrial vertebrates that can slide across the surface of the eyeball. It is reduced to a vestigial semilunar fold in human beings.

Nipple [nip′l] (Old English *neb* = beak, nose).
A papilla that bears the openings of the ducts from the mammary gland.

Nostril [nos′tril] (Old English *nosus* = nose + *thyrl* = hole).
The opening into the nasal cavity from the body surface (external nostril, naris) or from the nasal cavity into the pharynx (internal nostril, choana).

Obturator foramen [ob′tū-rā-tŏr] (L. *obturo*, p. p. *obturatus* = to close by stopping up).
A large foramen in the mammalian pelvic girdle. The obturator muscles arise from the periphery of the foramen and close it.

Occipital [ok-sip′i-tăl] (L. *occiput* = back of the head).
Pertaining to the back of the head or skull.

Occipital condyle [kon′dīl] (Gr. *kondylos* = condyle, knuckle).
One of a pair of enlargements on the occipital bone on each side of the foramen magnum of amphibians and mammals, or a single enlargement ventral to the foramen magnum in most other

vertebrates. They articulate with the cranial articular surface of the atlas.

Ocular [ok′yū-lăr] (L. *oculus* = eye).
Pertaining to the eye, e.g., the extrinsic ocular muscles that move the eyeball.

Oculomotor nerve [ok′yū-lō-mō-tŏr].
The third cranial nerve. It carries motor nerve fibers to most of the extrinsic muscles of the eyeball and autonomic nerve fibers into the eyeball.

Olecranon [ō-lek′ră-non] (Gr. = the tip of the elbow).
The proximal end of the ulna. It extends behind the elbow joint in mammals.

Olfactory [ol-fak′tŏ-rē] (L. *olfacio*, p. p. *olfactus* = to smell).
Pertaining to the parts of the nose involved in smelling.

Olfactory nerve.
The first cranial nerve, which consists of neurons returning from the nose to the olfactory bulb of the brain.

Omentum [ō-men′tŭm] (L. = fatty membrane).
The peritoneal membrane, often containing a great deal of fat. It extends between the body wall and stomach (i.e., greater omentum) or between the stomach and the liver and duodenum (i.e., lesser omentum).

Omo- [ō′mō] (Gr. *omos* = shoulder).
A combining form pertaining to the shoulder, e.g., omotransversarius muscle.

Oocyte [ō′ō-sīt] (Gr. *oon* = egg + *kytos* = cell).
An early stage in the development of the egg. The first meiotic division of the primary oocyte produces the secondary oocyte and a polar body.

Oogonium [ō-ō-gō′nē-ŭm] (Gr. *gone* = generation).
A very early stage in the development of an egg. It enlarges to become the primary oocyte.

Ootid [ō-ō-tid] (Gr. *ootidium* = a small egg).
The nearly mature egg, or ovum, after the second meiotic division has been initiated. In mammals, fertilization initiates the completion of this division.

Optic [op′tik] (Gr. *optikos* = pertaining to the eye).
Pertaining to the parts of the eye involved in vision.

Optic chiasma [kī-az′ma] (Gr. = cross, from the Greek letter *chi* = χ).
The complete or partial decussation of the optic nerve fibers on the ventral surface of the diencephalon.

Optic disk.
A disk-shaped area on the retina to which the optic nerve attaches. It lacks photoreceptors; also called the "blind spot."

Optic lobe.
One of a pair of lobes on the dorsal surface of the mesencephalon in nonmammalian vertebrates. It is a major integrating center in these vertebrates.

Optic nerve.
The second cranial nerve. It carries impulses from the retina to the brain.

Optic tract.
The neuronal tract leading from the optic chiasma to the thalamus, or optic lobes, or both.

Oral cavity [ōr′ăl] (L. *os*, gen. *oris* = mouth).
The mouth cavity; also called the buccal cavity.

Orbit [or′bit] (L. *orbis*, gen. *orbitis* = circle).
A circular cavity on the side of the skull. It lodges the eyeball.

Origin of a muscle.
The point of attachment of a muscle that tends to remain in a fixed position when the muscle contracts; the proximal end of a limb muscle.

Ossicle [os′i-kl] (L. *ossiculum* = small bone).
Any small bone, such as one of the auditory ossicles.

Osteocyte [os′tē-ō-sīt] (Gr. *osteon* = bone + *kytos* = cell).
A mature bone cell surrounded by the matrix it has produced.

Osteon [os′tē-on].
A cylindrical, microscopic unit of bone consisting of concentric layers of bone matrix surrounding a central canal that contains blood and lymph vessels. Also called a *Haversian system.*

Ostium [os′tē-ŭm] (L. = entrance, mouth).
The entrance to an organ, such as the ostium tubae of the uterine tube.

Otic capsule [ō′tik] (Gr. *otikos* = pertaining to the ear).
The part of the skull surrounding the inner ear.

Otolith [ō′tō-lith] (Gr. *oto-* = ear + *lithos* = stone).
Calcareous granules within sacs of the inner ear that stimulate sensory cells over which they move when an animal moves or changes position.

Oval window.
See **Fenestra vestibuli.**

Ovarian follicle [ō-var′ē-an] (L. *ovarium* = ovary + *folliculus* = little bag).
A group of epithelial and connective tissue cells in the ovary. It surrounds and nourishes the developing egg. It is also an endocrine gland whose primary hormone is estrogen.

Ovary [ō′vă-rē].
One of a pair of female reproductive organs containing the ovarian follicles and eggs.

Oviduct [ō′vi-dŭkt] (L. *ovum* = egg + *ducere*, p. p. *ductus* = to lead).

The passage in females of nonmammalian vertebrates. It transports eggs from the coelom to the cloaca.

Oviparous [ō-vip′ă-rŭs] (L. *pario* = to bear).
To bear eggs. A pattern of reproduction found in frogs and many nonmammalian vertebrates that lay eggs. The embryos develop outside the body of the mother.

Ovisac [ō′vi-sak].
An enlargement at the caudal end of the oviduct in which eggs accumulate before they are released.

Ovulation [ov′yū-lā-shŭn].
The release of mature egg cells from the ovarian follicles and ovary into the coelomic cavity (in the frog), ovarian bursa (in the rat and pig), or infundibulum (in the human female and many other mammals).

Ovum [ō′vum].
The mature egg cell.

Palate [pal′ăt] (L. *palatum* = palate).
The roof of the mouth. See also **Secondary palate.**

Pampiniform plexus [pam-pin′i-fōrm] (L. *pampinus* = tendril + forma = shape).
A network of veins in mammals entwining the testicular artery as it approaches the testis.

Pancreas [pan′krē-as] (Gr. *pan* = all + *kreas* = flesh).
A large glandular outgrowth of the duodenum. It secretes many digestive enzymes. It also contains the endocrine pancreatic islets of Langerhans.

Pancreatic islets.
Small clusters of endocrine cells in the pancreas. They produce hormones that regulate sugar metabolism; also called the *islets of Langerhans.*

Papilla [pă-pil′ă] (L. = nipple).
A small conical protuberance.

Papilla amphibiorum.
The part of the amphibian inner ear receptive to low-frequency sound waves.

Paradidymis [par′ă-did′i-mis] (Gr. *para* = beside + didymoi = testis).
A small group of vestigial kidney tubules occurring in male mammals and located beside the epididymis.

Paraflocculus [par-ă-flok′yū-lŭs] (L. *flocculus* = a small tuft of wool).
A small lobe on the lateral surface of the cerebellum. It is conspicuous in the rat.

Parasympathetic [par-ă-sim-pa-thet′ik] (Gr. *syn* = with + pathos = feeling).
Pertaining to the parasympathetic part of the autonomic nervous system. In mammals, preganglionic parasympathetic neurons leave the central nervous system through certain cranial nerves and sacral spinal nerves. The parasympathetic system stimulates metabolic processes that absorb and store energy.

Parathyroid gland [par-ă-thī′royd] (Gr. *thyreos* = oblong shield + *eidos* = shape).
One of several endocrine glands of a terrestrial vertebrate. It is located dorsal to, near, or entwined with the thyroid gland. Its hormone, called parathormone, helps to regulate calcium and phosphate metabolism.

Parietal [pă-rī′ĕ-tăl] (L. *paries,* gen. *parietis* = wall).
Pertaining to the wall of some structure, e.g., parietal bone, parietal peritoneum.

Parotid gland [pă-rot′id] (Gr. *para* = beside + otikos = pertaining to the ear).
A large mammalian salivary gland located ventral to the external ear.

Patella [pa-tel′ă] (L. = small plate).
The mammalian kneecap, composed of bone. It is a sesamoid bone within the patellar tendon of the quadriceps muscle.

Pectoral [pek′tŏ-răl] (L. *pectoralis* = pertaining to the chest).
Pertaining to the chest, e.g., pectoral muscles, pectoral appendage.

Pelvic [pel′vik] (L. *pelvis* = basin).
Pertaining to basin-shaped structures, such as the pelvic girdle, or renal pelvis.

Penis [pē′nis] (L. = tail, penis).
The male copulatory organ of most amniotes.

Pericardial cavity [per-i-kar′dē-ăl] (Gr. *peri* = around + kardia = heart).
The portion of the coelom that surrounds the heart.

Pericardium.
The serosa that covers the surface of the heart (i.e., visceral pericardium) and forms part of the pericardial wall (i.e., parietal pericardium).

Periosteum [per-ē-os′tē-ŭm] (Gr. *osteon* = bone).
The vascularized and innervated connective tissue covering a living bone.

Peritoneal cavity [per′i-tō-nē-al] (Gr. *peritonaion* = to stretch over).
The portion of the mammalian coelom that houses the abdominal viscera.

Peritoneum.
The serosa that covers the abdominal visceral organs (i. e., visceral peritoneum) and lines the peritoneal cavity (i.e., parietal peritoneum).

Peroneus muscle [per-ō-nē′ŭs] (Gr. *perone* = pin, fibula).
One or more muscles located on the lateral side of the shin over the fibula and extending into the foot.

Phalanges [fă-lan′jēz] (Gr. = battle lines of soldiers).
Bones of the digits that extend beyond the palm of the hand or sole of the foot.

Pharynx [far′ingks] (Gr. = throat).
That part of the digestive tract that lies between the oral cavity and the esophagus; the crossing place of the digestive and respiratory tracts. Embryonically, lungs develop as outgrowths from the pharyngeal floor, and pharyngeal pouches develop from the lateral walls of the pharynx.

Phrenic [fren′ik] (Gr. *phren* = diaphragm).
Pertaining to the diaphragm, e.g., phrenic nerve, phrenic artery.

Pia mater [pī′ă mā′ter] (L. = tender mother).
The delicate, vascularized layer of connective tissue that invests the brain and spinal cord. It is the innermost of the three mammalian meninges.

Pineal gland [pin′ē-ăl] (L. *pineus* = pertaining to pine, from *pinus* = pine tree).
An endocrine gland that produces melatonin, especially under dark conditions. The functions of melatonin are not fully understood, but it has been implicated in adjusting physiological processes to diurnal and seasonal cycles.

Piriformis muscle [pir′i-fōrm-is] (L. *pirum* = pear + *forma* = shape).
A small, triangular muscle located on the medial side of the proximal end of the thigh. It is pear-shaped in human beings.

Pituitary gland.
See **Hypophysis.**

Placenta [plă-sen′tă] (L. = flat cake).
The apposition or union of parts of the uterine lining and fetal extraembryonic membranes through which food, respiratory gases, and waste products are exchanged between the mother and the fetus.

Plantaris muscle [plan′tār-is] (L. = pertaining to the sole of the foot).
A large muscle on the caudal surface of the crus of nonmammalian vertebrates. Its tendon runs onto the sole of the foot and extends the foot.

Platelet [plāt′let] (English, *small plate*).
An irregularly shaped fragment of the cytoplasm of a megakaryocyte in the blood. It is involved in blood clotting.

Platysma muscle [plă-tiz′mă] (Gr. *platys* = flat, broad + *-ma* = suffix indicating result of an action, e.g., of making something flat and broad).
Thin sheet of muscle underlying the skin of the neck in mammals.

Pleura [plūr′ă] (Gr. = side, rib).
The serosa that covers the lungs (i.e., visceral pleura) and lines the pleural cavities (i.e., parietal pleura).

Pleural cavities.
The paired coelomic spaces that enclose the lungs in mammals.

Pleuroperitoneal cavity [plūr-ō-per-i-tō-nē′al].
The combined potential peritoneal and pleural cavities in anamniotes. It contains the abdominal viscera and, if present, the lungs.

Plexus [plek′sŭs] (L. = network, braid).
A network of blood vessels or nerves, e.g., choroid plexus, brachial plexus.

Polar body [pō′lăr] (L. *polaris* = pole).
A small body located near the animal pole of a developing egg cell. It results from the unequal division of the egg during the first and second meiotic division.

Pollex [pol′eks] (Gr. = thumb).
The thumb.

Pons [ponz] (L. = bridge).
The ventral part of the mammalian metencephalon. Its most conspicuous surface feature is a bridgelike tract of transverse neuron fibers.

Popliteal [pop-lit′ē-ăl] (L. *poples*, gen. *poplitis* = knee joint).
The depression behind the mammalian knee joint.

Portal vein [pōr′tăl] (L. *porta* = gate, door).
A vein that carries blood from the capillaries of one organ to the capillaries of another organ rather than to the heart, e.g., the hepatic portal vein.

Posterior [pos-tēr′ē-ōr] (L. *poster* = after, following + *ior* = suffix indicating the comparative form of an adjective, e.g., more behind).
A direction toward the back of a human being. It is sometimes used for the tail end of a quadruped, but *caudal* is a more appropriate term.

Posterior chamber.
The space within the eyeball between the iris and lens. It is filled with aqueous humor.

Postganglionic motor neuron [post′gang-glē-on′ik] (Gr. *ganglion* = small tumor, swelling).
A neuron of the autonomic nervous system whose cell body lies in a ganglion in the peripheral nervous system.

Preganglionic motor neuron [prē′gang-glē-on′ik] (L. *pre-* = in front of).
A neuron of the autonomic nervous system whose cell body lies within the central nervous system and whose axon synapses with a postganglionic neuron.

Premolar tooth [prē-mō′lăr] (L. *molaris* = millstone).
   One of several mammalian teeth that lie in front of the molar teeth and behind the canine tooth. They usually are adapted for a combination of cutting and grinding.

Prepuce [prē′pūs] (L. *praeputium* = foreskin).
   The foreskin of a male mammal. It covers the glans penis.

Prosencephalon [prōs-en-sef′ă-lon] (Gr. *pro-* = before + *enkephalos* = brain).
   The embryonic forebrain. It gives rise to the telencephalon and diencephalon.

Prostate [pros′tāt] (Gr. *prostates* = one who stands before).
   An accessory genital gland of male mammals. It surrounds the urethra just before the urinary bladder and secretes much of the seminal fluid.

Protraction [prō-trak′shŭn] (L. *traho*, p. p. *tractus* = to pull).
   Muscle action that moves the entire appendage of a quadruped forward.

Proximal [prok′si-măl] (L. *proximus* = nearest).
   The end of a structure nearest its origin, e.g., the end of a limb that is closest to the girdle.

Pterygoideus muscle [ter′i-goyd-e-us] (Gr. *pteryx, pteryg-* = wing + *eidos* = resembling).
   A jaw-closing muscle that arises from the pterygoid bone (frog) or pterygoid process on the underside of the skull (mammals) and inserts on the medial side of the lower jaw.

Pubis [pyū′bis] (L. = genital hair).
   The cranioventral bone of the pelvic girdle of terrestrial vertebrates.

Pudendal [pyū-den′dăl] (L. *pudendum* = external genitals, from *pudendo* = to feel ashamed).
   Pertaining to the region of the external genitals, e.g., pudendal artery.

Pulmonary [pŭl′mō-nār-ē] (L. *pulmo* = lung).
   Pertaining to the lungs.

Pulmonary trunk.
   The mammalian arterial trunk that leaves the right ventricle and soon divides into the two pulmonary arteries going to the lungs.

Pulmonary valve.
   A set of three semilunar folds in the base of the pulmonary trunk. It prevents the backflow of blood into the right ventricle.

Pupil [pyū′pĭl] (L. *pupilla* = pupil).
   The central opening in the iris through which light enters the eyeball.

Pylorus [pī-lōr′ŭs] (Gr. *pyloros* = gatekeeper).
   The caudal end of the stomach, which contains a sphincter muscle.

Pyramidal system [pi-ram′i-dal] (Gr. *pyramis* = pyramid).
   The direct neuronal motor pathway in mammals, from pyramid-shaped cell bodies in the cerebrum to motor neurons in the brain and spinal cord.

Quadrate cartilage [kwah′drāt] (L. *quadratus* = square).
   A cartilage at the caudal end of the upper jaw of a frog with which the mandibular cartilage of the lower jaw articulates. It represents the dorsal half of the first visceral arch of ancestral fishes.

Radius [rā′dē-ŭs] (L. = ray, spoke).
   The bone of the forearm that rotates around the ulna in most terrestrial vertebrates. It lies on the lateral side of the forearm when the palm is supine.

Ramus [rā′mŭs] (L. = branch).
   A branch of a nerve or a blood vessel, also part of the mandible.

Rectum [rek′tŭm] (L. *rectus* = straight).
   The caudal end of the large intestine of a mammal.

Renal [rē′năl] (L. *ren* = kidney).
   Pertaining to the kidney, e.g., renal artery.

Renal portal system.
   A system of veins that drains the capillaries of the hind legs and tail of most nonmammalian vertebrates and leads to capillaries around the kidney tubules.

Renal tubule.
   A kidney tubule. It consists of a glomerular capsule and a tubule; the tubular part of a nephron.

Rete testis [rē′tē tes′tis] (L. = net + *testis* = testicle).
   A network of small passages in the mammalian testis between the seminiferous tubules and the epididymis. They form a visible cord in the rat.

Retina [ret′i-nă].
   The innermost layer of the eyeball. It contains pigment, receptive rods and cones, and associated neurons.

Retraction [rē-trak′shŭn] (L. *re* = backward + *tractio* = a pull).
   Muscle action that pulls the entire appendage of a quadruped caudad.

Rhinal [rī′năl] (Gr. *rhis*, gen. *rhinos* = nose).
   Pertaining to the nose.

Rhinal sulcus.
   The furrow that separates the portion of the cerebrum dealing with olfaction from other areas of the brain.

Rhinencephalon [rī′nen-sef′ă-lon] (Gr. *enkephalos* = brain).
   The primary olfactory portion of the cerebrum.

Rhombencephalon [rom-ben-sef′ă-lon] (Gr. *rhombus* = an oblique-angled equilateral parallelogram + *enkephalos* = brain).

The embryonic hindbrain. It develops into the metencephalon and myelencephalon.

Rodentia [rō-den′shē-ă] (L. *rodo*, pres. p. *rodens* = to gnaw).
The mammalian order, including the rat and other gnawing animals, that is characterized by two enlarged incisor teeth in the upper and lower jaws.

Rostral [ros′trăl] (L. *rostrum* = beak).
A direction toward the front of the head. This term is used for structures within the head.

Round ligament.
A round cord; specifically, a connective tissue cord that crosses the broad ligament in female mammals and connects the ovary with the body wall at the groin. It corresponds to the male gubernaculum.

Round window.
See **Fenestra cochleae.**

Sacculus [sak′yū-lŭs] (L. small sac).
The most ventral chamber of the inner ear. It contains an otolith and functions as a sensor of equilibrium.

Sacral vertebrae, sacrum [sa′krŭm] (L. = sacred).
The vertebrae, or fusion of two or more vertebrae and their ribs, with which the pelvis articulates.

Sagittal [saj′i-tăl] (L. *sagitta* = arrow).
A plane of the body that passes through the sagittal suture (between the parietal bones of the skull), e.g., a median, longitudinal plane passing from dorsal to ventral.

Salivary gland [sal′i-vār-ē] (L. *saliva* = saliva).
One of several glands that secrete saliva. Major ones in mammals are the parotid, mandibular, and sublingual glands.

Sarcolemma [sar′kō-lem-ă] (Gr. *sarx*, gen,. *sarkos* = flesh + *lemma* = husk).
The cell membrane of a muscle cell.

Sartorius muscle [sar-tōr′ē-ŭs] (L. *sartor* = tailor).
A narrow, diagonal muscle on the medial surface of the thigh. It is especially well developed in people who sit cross-legged on the floor, such as tailors in preindustrial times, because it is used to rise.

Scapula [skap′yū-lă] (L. = shoulder blade).
The shoulder blade, or the element of the pectoral girdle that extends dorsally on the back.

Sclera [sklēr′a] (Gr. *skleros* = hard).
The opaque (white) portion of the fibrous tunic of the eyeball. Together with the cornea, it forms the outer wall of the eyeball.

Scrotum [skrō′tŭm] (L. = pouch).
The cutaneous sac that encases the paired mammalian testes when the testes are descended.

Sebaceous gland [sē-bā′shŭs] (L. *sebum* = tallow).
A gland in mammalian skin. It produces an oily or waxy secretion, which is usually discharged into a hair follicle.

Secondary palate.
The palate of mammals that separates the food and air passages. It consists of the hard palate, which separates the oral from the nasal cavities, and the fleshy soft palate, which separates the oral pharynx from the nasal pharynx.

Semicircular duct [sem′ē-sir-kyū-lăr] (L. *semi* = prefix denoting one-half).
One of three semicircular ducts in the inner ear. Each lies at a right angle to the others and detects changes in angular acceleration generated by turns of the head.

Seminal fluid [sem′i-năl] (L. *semen* = seed).
The fluid secreted by male reproductive ducts and accessory genital glands.

Seminal vesicle.
See **Vesicular gland.**

Seminiferous tubules [sem-i-nif′er-ŭs] (L. *semen*, gen. *seminis* = seed + *fero* = to carry).
Tubules within the testis in which sperm cells are produced.

Septum [sep′tum] (L. = partition).
A partition.

Septum pellucidum [pe-lū′si-dum] (L. *pellucidus* = clear, transparent).
A thin, vertical septum of nervous tissue in the brain ventral to the corpus callosum. It forms the medial wall of each lateral ventricle.

Serosa [se-rō′să] (L. *serosus* = watery).
The coelomic epithelium and underlying connective tissue that line body cavities and cover visceral organs, e.g., peritoneum, pleura, pericardium, and tunica vaginalis.

Serratus muscle [ser-āt′us] (L. = toothed like a saw).
A muscle with a serrated or saw-toothed border.

Sertoli cell (*Enrico Sertoli*, Italian histologist, 1842–1910).
Large epithelial cell in the seminiferous tubules. It plays a role in the maturation of the sperm cells.

Sesamoid bone [ses′ă-moyd] (Gr. *sesamon* = sesame seed + *eidos* = resembling).
A bone that develops in the tendon of some muscles near their insertion and facilitates the movement of the tendon across a joint, or changes the direction of the force of a muscle, e.g., the patella and the pisiform.

Sinus [sī-nŭs] (L. = a cavity).
A cavity or space within an organ.

Sinus venosus [vē-nō′sus] (L. *venosus* = pertaining to a vein).

Glossary

RHO–SIN

The most caudal chamber of the heart of anamniotes and some reptiles. In frogs, it receives blood from the body and leads to the right atrium.

Skin.
> See **Integument.**

Skull [skŭl] (Old English *skulle* = a bowl).
> The group of bones and cartilages that encase the brain and major sense organs and form the upper jaw and face. The lower jaw sometimes is considered to be a part separate from the skull.

Soft palate.
> The fleshy partition in mammals between the oral pharynx and nasal pharynx.

Somatic [sō-mat′ik] (Gr. *somatikos* = bodily).
> Pertaining to the body wall and appendages, as opposed to the internal (visceral) organs, e.g., somatic muscles, somatic skeleton.

Sperm cell [sperm] (Gr. *sperma* = seed, sperm cell).
> One of the mature male gametes; also called spermatozoa.

Spermatid [sper′mă-tid] (Gr. *idion,* a diminutive ending).
> A cell resulting from the second meiotic division of a secondary spermatocyte. It develops into a sperm cell.

Spermatocyte [sper′mă-tō-sīt] (Gr. *kytos* = cell).
> The primary spermatocyte results from the growth and division of a spermatogonium, and the secondary spermatocyte results from the first meiotic division of the primary spermatocyte.

Spermatogonium [sper′mă-tō-gō-nē-ŭm] (Gr. *gone* = generation).
> The stem cell that multiplies mitotically and gives rise to sperm cell-forming cells.

Spermatozoa.
> See **Sperm cell.**

Spinal [spī′năl] (L. *spina* = spine, thorn).
> Pertaining to a spine-shaped structure, often the spine or vertebral column, e.g., spinal cord, spinal nerve.

Spinous process.
> The dorsal, spinelike process of the vertebral arch.

Splanchnic [splangk′nik] (Gr. *splanchnon* = gut, viscus).
> Pertaining to structures that supply the gut or visceral organs, such as the splanchnic nerve.

Spleen [splēn] (Gr. *splen* = spleen).
> A large lymphoid organ located near the left side of the stomach. In various vertebrates and at different times in their life cycle, it produces, stores, or disassembles senescent red blood cells.

Splenius muscle [splē′nē-ŭs] (Gr. *splenion* = bandage).
> A thin, triangular sheet of muscle located on the back of the neck of mammals deep to part of the rhomboideus muscle.

Stapes [stā′pēz] (L. = stirrup).
> The single auditory ossicle of nonmammalian vertebrates, and the innermost of the three auditory ossicles in mammals.

Statoacoustic nerve [stat′ō-ă-kū′stik] (Gr. *statos* = standing still + *akoustikos* = pertaining to hearing).
> A name often used for the eighth cranial nerve in nonmammalian vertebrates. It is called the vestibulocochlear nerve in mammals.

Sternum [ster′nŭm] (Gr. *sternon* = chest).
> The breastbone of terrestrial vertebrates.

Stomach [stŭm′ŭk] (Gr. *stomakos* = stomach).
> Saclike part of the digestive tract lying between the esophagus and intestine in which food is temporarily stored and digestion usually is initiated.

Stratum [strat′ŭm] (L. = layer).
> A layer of tissue, such as the stratum corneum of the skin.

Styloid process [stī′loyd] (Gr. *stylos* = pillar + *eidos* = resembling).
> A slender process on the ventral surface of the skull of many mammals. The hyoid apparatus attaches to it.

Subclavian [sŭb-klā′vē-an] (L. *sub-* = beneath + *clavis* = key).
> Pertains to a position beneath the clavicle, e.g., the subclavian artery.

Sublingual gland [sŭb-ling′gwăl] (L. *lingua* = tongue).
> A salivary gland of mammals. It lies beneath the tongue.

Submucosa [sŭb-mŭ-kō′să] (L. *mucosus* = mucous).
> A layer of vascularized connective tissue in the wall of the digestive or respiratory tract. It underlies the mucosa.

Sulcus [sŭl′kŭs] (L. = groove).
> A groove on the surface of an organ, e.g., the sulci on the surface of the cerebrum of many mammals.

Superior [sū-pēr′ē-ōr] (L. *super* = above + *-ior* = suffix indicating the comparative form of an adjective, e.g., more above).
> A direction toward the head of a human being.

Suprarenal gland [sū-pră-rē′năl].
> See **Adrenal gland.**

Suture [sū′chūr] (L. = seam).
> An essentially immovable joint in which the bones are separated by connective tissue, e.g., many of the joints between bones of the skull.

Sympathetic nervous system [sim-pă-thet′ik] (Gr. *syn* = with + *pathos* = feeling).
> The part of the autonomic nervous system that, in mammals, leaves the central nervous system from

the thoracic and lumbar parts of the spinal cord. Its activity helps an animal adjust to stress by promoting physiological processes that mobilize the energy available to the tissues.

Symphysis [sim′fi-sis] (Gr. *physis* = a growth).
A joint or fusion between bones that permits limited movement by the deformation of fibrocartilage between them. It occurs in the midline of the body, e.g., the pelvic symphysis, mandibular symphysis.

Systemic [sis-tem′ik] (Gr. *systema* = whole).
Pertaining to the body as a whole rather than to a specific part, e.g., the systemic circulation as opposed to the pulmonary circulation.

Talus [tā′lus] (L. = ankle bone).
The proximal tarsal bone of mammals. It articulates with the tibia.

Tapetum lucidum [tă-pē′tŭm lū′sid-ŭm] (L. = carpet + *lucidus* = shining).
A layer within or behind the retina of some vertebrates. It reflects light back onto the photoreceptive cells.

Tarsal [tar′săl] (Gr. *tarsos* = ankle).
One of the small bones of the ankle.

Tela choroidea [tē′la kōr-oy′dē-a] (L. *tela* = web, Gr. *chorion* = membrane encasing the fetus + *eidos* = resembling).
A thin membrane forming the roof or part of the wall of some ventricles of the brain. It is composed of the ependymal epithelium and the pia mater.

Telencephalon [tel-en-sef′ă-lon] (Gr. *telos* = end + *enkephalos* = brain).
The most rostral of the five brain regions. It includes the olfactory lobes and cerebrum.

Temporal [tem′pŏ-răl] (L. *tempus*, gen. *temporis* = time).
Pertaining to the temporal region of the skull, so called because the hair in this region is the first to become gray in human beings.

Temporal fossa.
A depression on the lateral surface of the mammalian skull caudal to the orbit and dorsal to the zygomatic arch. It lodges the temporal jaw muscle.

Tendon [ten′dŏn] (L. *tendo* = to stretch).
A band of dense connective tissue attaching a muscle to a bone, or sometimes to another muscle.

Tendon of Achilles (*Achilles*, the hero of Homer's *Iliad* was said to be invulnerable except for this tendon).
The tendon that extends from the large muscle mass on the caudal surface of the crus to the calcaneus. These muscles are powerful extensors of the foot.

Tentorium [ten-tō′rē-um] (L. = tent, from *tendo*, p. p. *tentum* = to stretch).
A septum of dura mater in mammals located between the cerebrum and cerebellum. It ossifies in some species.

Teres [ter′ēz] (L. = rounded, smooth).
Descriptive of a round structure, such as the teres muscles.

Tertiary follicle.
A mature follicle in the ovary; also called a *Graafian follicle.*

Testis [tes′tis] (L. = witness, in ancient Rome necessarily an adult male).
The male gonad. It produces sperm cells and the hormone testosterone.

Tetrapod [tet′ră-pod] (Gr. *tetra* = four + *pous*, gen. *podos* = foot).
A collective term for terrestrial vertebrates. They have four feet unless some have been secondarily lost or modified.

Thalamus [thal′ă-mus] (Gr. *thalamos* = inner chamber, bedroom).
The lateral walls of the diencephalon. It is an important center between the cerebrum and other parts of the brain.

Theca [thē′kă] (Gr. *theke* = box, case).
A case or covering, such as the group of connective tissue cells that form the outer layer of an ovarian follicle.

Thoracolumbar fascia [thōr′ă-kō-lŭm′bar].
A part of the deep fascia on the dorsal surface of the thoracic and lumbar regions. It is associated with the muscles of the back and abdominal wall.

Thorax [thō′raks] (Gr. = chest).
The region of the mammalian body encased by the ribs and sternum.

Thymus [thī′mŭs] (Gr. *thymos* = thyme, thymus; so called because of its resemblance to a bunch of thyme).
A lymphoid organ in the ventral part of the neck and thorax. It is essential for the maturation of T-lymphocytes and probably other parts of the immune system. The gland is best developed in young individuals and atrophies later in life.

Thyroid gland [thī′royd] (Gr. *thyreos* = oblong shield + *eidos* = resembling).
An endocrine gland that is usually located near the cranial end of the trachea, but lies over the thyroid cartilage of the larynx in human beings (whence its name). Its hormones increase the rate of metabolism.

Tibia [tib′ē-ă] (L. = the large shinbone).
The large bone on the medial side of the lower leg.

Tissue [tish′ū] (Old French *tissu* = cloth).
An aggregation of cells that together perform a common function.

Tongue [tŭng] (Old English *tunge* = tongue).
A muscular organ in the floor of the oral cavity that often helps gather food and manipulates it within the mouth cavity.

Tonsil [ton′sil] (L. *tonsilla* = tonsil).
One of the lymphoid organs that develop in the wall of the mammalian pharynx.

Trachea [trā′kē-ă] (Gr. *tracheia* = rough artery).
The respiratory tube between the larynx and the bronchi.

Tract [trakt] (L. *traho*, p.p. *tractus* = to pull).
(1) A linear group of organs having a similar function, e.g., the digestive tract. (2) A group of axons of similar function traveling together in the central nervous system.

Transverse [trans-vers′] (L. *transversus* = transverse).
A plane of the body crossing its longitudinal axis at right angles.

Transverse process.
A process of a vertebra, that lies in the transverse plane. Either the tubercle of a rib articulates with it, or an embryonic rib becomes incorporated in it and serves as an attachment site for muscles.

Trapezoid body [trap′ĕ-zōyd] (Gr. *trapezoides* = resembling a trapezium).
An acoustic commissure at the rostral end of the ventral surface of the mammalian medulla oblongata.

Trigeminal nerve [trī-jem′i-năl] (L. *trigeminus* = threefold).
The fifth cranial nerve. It has three branches in mammals. It innervates the jaw muscles and returns sensory fibers from the surface of the head and oral cavity, except for taste buds.

Trochanter [trō-kan′ter] (Gr. = a runner).
One of the processes on the proximal end of the femur to which certain pelvic and thigh muscles attach.

Trochlear nerve [trok′lē-ăr] (L. *trochlea* = pulley).
The fourth cranial nerve. It innervates one of the extrinsic muscles of the eye, which, in mammals, passes through a connective tissue pulley before inserting on the eyeball.

Truncus arteriosus [trŭng′kŭs ar-ter′ē-ō-sus] (L. = trunk, stem).
One of two arterial trunks in the frog leading from the cranial end of the heart to arterial arches supplying the skin and lungs, head, and body.

Tunic [tū′nik] (L. *tunica* = a coat, covering).
Descriptive of a layer of an organ, such as one of the layers of the eyeball.

Tunica albuginea [tū′ni-kă al-byū-jin′ē-ă] (L. *albugineus*, from *albugo* = white spot).
A white, fibrous capsule, such as the one forming the wall of the testis and sending septa into the testis.

Tympanic [tim-pan′ik] (L. *tympanum* = drum).
Pertaining to the middle ear.

Tympanic cavity.
The cavity of the middle ear that lies between the tympanic membrane and the inner ear within the otic capsule. One or three auditory ossicles traverse it, and the auditory tube connects it with the pharynx.

Tympanic membrane.
The eardrum.

Ulna [ŭl′nă] (L. = elbow bone).
One of the bones of the antebrachium of terrestrial vertebrates. It extends behind the elbow in mammals and lies on the medial side of the antebrachium when the hand is supine.

Umbilical [ŭm-bil′i-kăl] (L. *umbilicus* = navel).
Pertaining to the navel, e.g., umbilical cord, umbilical artery.

Ureter [yū-rē′ter] (Gr. *oureter* = ureter, from *ouron* = urine).
The duct in amniotes that carries urine from the kidney to the urinary bladder.

Urethra [yū-rē′thră] (Gr. *ourethra* = urethra).
The duct that carries urine from the urinary bladder to the cloaca or to the outside of the body in amniotes; part of it also carries spinal fluid and sperm cells in males.

Urinary bladder [yūr′i-nār-ē] (Gr. *ouron* = urine).
A saccular organ in terrestrial vertebrates in which urine from the kidney accumulates before being discharged from the body.

Urogenital [yū′rō′jen-i-tăl] (L. *genitalis* = creative, fruitful).
A combining term for the urinary and genital systems, which share many ducts.

Urostyle [yū′rō′stīl] (Gr. *oura* = tail + *stylos* = a pillar, peg).
The rodlike, caudal part of the vertebral column of a frog. It is composed of several fused caudal vertebrae.

Uterine tube [yū′ter-in] (L. *uterus* = womb).
One of a pair of narrow tubes in mammals. It extends from the vicinity of the ovary to the uterus and carries fertilized eggs to the uterus. Also called the *Fallopian tube.*

Uterus [yū′ter-ŭs].
The organ in females in which embryos develop in live-bearing species. It develops from part of the oviduct.

Utriculus [yū′trik′yū-lŭs] (L. = small sac).
The upper chamber of the inner ear to which the semicircular ducts attach.

Vagina [vă-jī′na] (L. = sheath).
The part of the mammalian female reproductive tract that receives the penis during copulation.

Vaginal cavity.
The part of the coelom that contains the testis of male mammals.

Vaginal vestibule [ves′ti-būl] (L. *vestibulum* = antechamber).
The passage or space into which the vagina and urethra enter in female mammals. It is very shallow in human beings but often forms a short canal in quadrupeds.

Vagus nerve [vā′gŭs] (L. = wandering).
The tenth cranial nerve. It carries motor fibers to muscles of the larynx and parasympathetic fibers to thoracic and abdominal organs; it returns sensory fibers from these parts of the body.

Vasa efferentia.
See **Efferent ductules.**

Vascular tunic.
The middle layer of the eyeball. It forms the choroid, ciliary body, and iris.

Vas deferens.
See **Ductus deferens.**

Vastus [vas′tus] (L. = great, large).
Descriptive of some thigh muscles, e.g., the vastus lateralis.

Vein [vān] (L. *vena* = vein).
A blood vessel that carries blood toward the heart. Usually the blood is low in oxygen content, but pulmonary veins from the lungs have blood with a high oxygen content.

Vena cava [vē′nă cā′vă] (L. = hollow vein).
One of the primary veins of frogs and amniotes. It leads directly to the heart.

Ventral [ven′tral] (L. *ventralis* = ventral, from *venter* = belly).
A direction toward the underside of a quadruped.

Ventricle [ven′tri-kl] (L. *ventriculus* = small belly).
(1) A chamber of the heart. It greatly increases blood pressure and sends the blood to the lungs or to the body. (2) One of the cavities within the brain.

Vermis [ver′mis] (L. = worm).
The "segmented" and wormlike median portion of the mammalian cerebellum.

Vertebra [ver′te-bră] (L. = vertebra, joint).
One of the units that make up the vertebral column.

Vertebral arch.
The arch of a vertebra that surrounds the spinal cord. It is also called a *neural arch.*

Vertebral body.
The main, supporting component of a vertebra. It lies ventral to the vertebral arch. Also called a vertebral centrum.

Vesicular gland [vĕ-sik′yū-lăr] (L. *vesicula* = small bladder).
One of the accessory genital glands of male mammals that contributes to the seminal fluid; also called the *seminal vesicle.*

Vestibulocochlear nerve [ves-tib′yū-lō-kok-lē-ăr] (L. *cochlea* = snail shell).
The eighth cranial nerve. It returns sensory fibers from the parts of the inner ear related to equilibrium (vestibular apparatus) and from the part monitoring sound detection (cochlea). Often called the statoacoustic nerve in anamniotes, in which a cochlea is absent.

Vibrissae [vī-bris′ē] (L. = vibrissae, from *vibro* = to quiver).
Long, tactile hairs on the snout of many mammals.

Villi [vil′i] (L. = shaggy hair).
Multicellular, but minute, often finger-shaped projections of an organ. They increase its surface area, e.g., the intestinal villi.

Visceral [vis′er-ăl] (L. *viscus*, pl. *viscera* = internal organs).
Pertaining to the inner part of the body as opposed to the body wall and appendages, e.g., visceral muscles, visceral skeleton.

Visceral arches.
The skeletal arches that develop in the wall of the pharynx and may contribute to the formation of the jaws (frog) and parts of the skull, hyoid, and larynx.

Vitreous body [vit′rē-ŭs] (L. *vitreus* = glassy).
The clear, viscous material in the eyeball between the lens and retina.

Viviparous [vī-vip′ă-rŭs] (L. *vivus* = living + *pario* = bearing).
Live-bearing. A pattern of reproduction found in most mammals and a few other vertebrates in which the young develop in a uterus and are born fully formed.

Vocal cords [vō′kăl] (L. *vocalis* = pertaining to voice).
Folds of mucous membrane within the larynx of frogs and mammals. Involved in the production of sounds.

Vomer bone [vō′mer] (L. = plowshare).
A bone in the roof of the mouth in frogs and in the floor of the nasal cavities in mammals. It is plowshare-shaped.

Vomeronasal organ [vō′mer-ō-nā-zăl].
An accessory olfactory organ located between the

palate and the nasal cavities of most terrestrial vertebrates. It is important in feeding and sexual behaviors. It is also called **Jacobson's organ.**

Vulva [vŭl′vă] (L. = covering).
The female external genitalia.

Wharton's jelly (*Thomas Wharton,* British anatomist and physician, 1614 - 1673).
The mucoid connective tissue of the umbilical cord.

White matter.
Tissue in the central nervous system that consists primarily of myelinated axons.

Wolffian duct (*Kaspar Friedrich Wolff,* German embryologist in Russia, 1733-1794).
See **Archinephric duct.**

Yolk sac [yōk] (Old English *geulca* = yolk, from *geolu* = yellow).
The yolk-containing sac attached to the ventral surface of the embryo in some fishes, reptiles, birds, and egg-laying mammals. It is reduced in viviparous mammals.

Zonule fibers [zō′nyūl] (L. *zonula* = small zone).
Delicate fibers extending between the ciliary body and the lens equator. These fibers transmit forces from the ciliary body to the lens.

Zygomatic arch [zī ′gō-mat-ik] (Gr. *zygoma* = bar, yoke).
The arch of bone beneath the orbit in a mammalian skull. It connects the facial and cranial regions.